David Sang

Cambridge IGCSE® Physics Workbook

CAMBRIDGE UNIVERSITY PRESS
Cambridge, New York, Melbourne, Madrid, Cape Town,
Singapore, São Paulo, Delhi, Mexico City

Cambridge University Press
The Edinburgh Building, Cambridge CB2 8RU, UK

www.cambridge.org
Information on this title: www.cambridge.org/9780521173582

First published 2010
5th printing 2012

Printed in the United Kingdom at the University Press, Cambridge

A catalogue record for this publication is available from the British Library

ISBN 978-0-521-17358-2 Paperback

Cover image: Fingers weave through optical fibers.

NOTICE TO TEACHERS

References to practical activities contained in these resources are provided 'as is' and
information provided is on the understanding that teachers and technicians
shall undertake a thorough and appropriate risk assessment before
undertaking any of the practical activities listed. Cambridge University Press makes
no warranties, representations or claims of any kind concerning the practical activities.
To the extent permitted by law, Cambridge University Press will not be liable
for any loss, injury, claim, liability or damage of any kind resulting from the
use of the practical activities.

Contents

Introduction

This book has been written to help you increase your understanding of the topics covered in your IGCSE Physics course. The exercises will give you opportunities for the following:

- practice in writing about the ideas that you are studying
- practice in solving numerical and other problems
- practice in thinking critically about experimental techniques and data
- practice in drawing and interpreting diagrams, including graphs.

Most of the exercises are somewhat different from examination questions. This is because they are designed to help you *develop* your knowledge, skills and understanding. (Examination questions are designed differently, to *test* what you know, understand and can do.)

Spaces have been left for you to write your answers. Some of the diagrams are incomplete, and your task will be to complete them.

Safety

A few practical exercises have been included. These could be carried out at home using simple materials that you are likely to have available to you. (There are many more practical activities on the CD-ROM that accompanies your textbook.)

While carrying out such experiments, it is your responsibility to think about your own safety, and the safety of others. If you work sensibly and assess any risks before starting, you should come to no harm. If you are in doubt, discuss what you are going to do with your teacher before you start.

Block 1
General physics

1 Making measurements

A definition to learn

density the ratio of mass to volume for a substance

An equation to learn and use

$$\text{density} = \frac{\text{mass}}{\text{volume}}$$

Exercise 1.1 Accurate measurements

To measure a length accurately, it is essential to have a careful technique. Special measuring instruments can also help.

a The diagram shows how a student attempted to measure the length of a piece of wire.

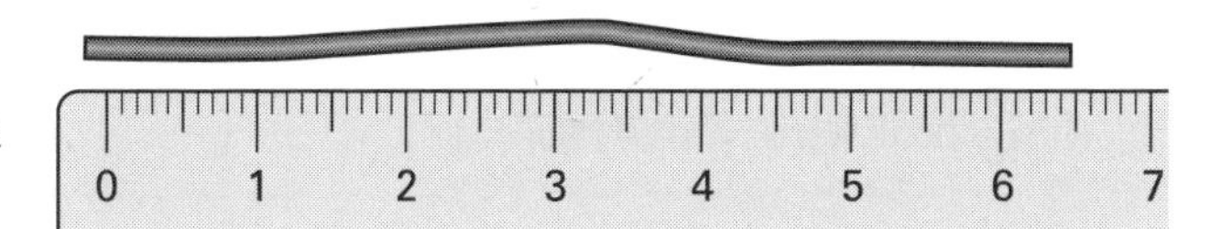

From the diagram, estimate the length of the wire. ....

b State **three** ways in which the student could have improved his technique for measuring the wire.

....

....

....

E

c The diagram shows a set of vernier callipers. Label the following parts of this measuring instrument:

vernier scale   main scale   jaws

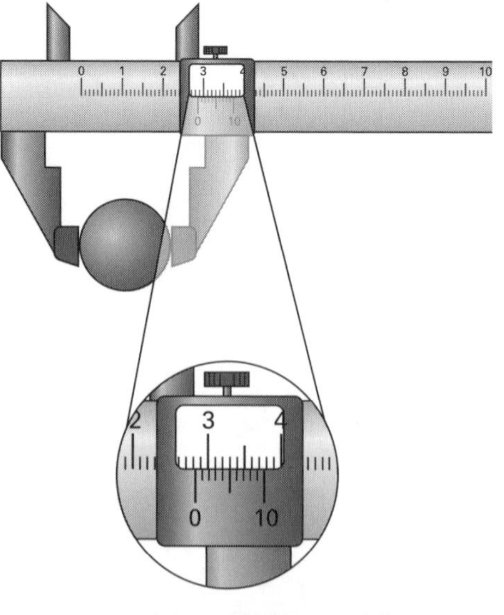

d Determine the diameter of the ball, as measured by the vernier callipers shown on the right.

....

e A micrometer screw gauge can be used to measure the thickness of a sheet of plastic. What value is shown in the diagram on the right?

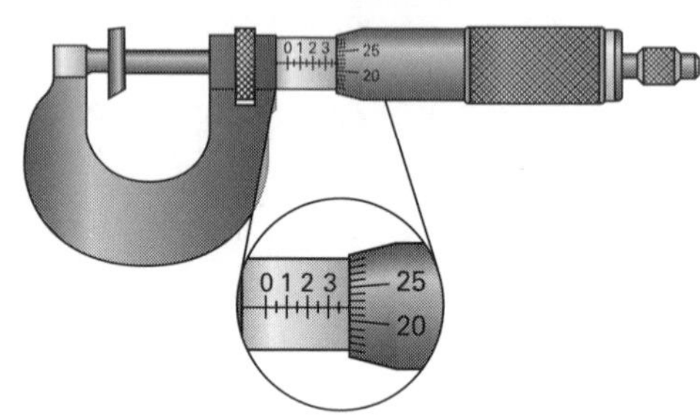

....

f During an experiment, a student made the measurements shown in the table below. In the second column, suggest the instrument that she used to make each measurement.

Measurement	Measuring instrument
length of wire = 20.4 cm	
thickness of wire = 4.24 mm	
thickness of wooden block = 17.5 mm	

Exercise 1.2 Density data

This exercise presents some data for you to interpret and use.

The table on the opposite page shows the densities of some solids and liquids. Two units are used, kg/m^3 and g/cm^3.

Material	State / type	Density / kg/m³	Density / g/cm³
water	liquid / non-metal	1000	1.000
ethanol	liquid / non-metal	800	0.800
olive oil	liquid / non-metal	920	
mercury	liquid / metal	13500	
ice	solid / non-metal	920	
diamond	solid / non-metal	3500	
cork	solid / non-metal	250	
chalk	solid / non-metal	2700	
iron	solid / metal	7900	
tungsten	solid / metal	19300	
aluminium	solid / metal	2700	
gold	solid / metal	19300	

a Complete the last column by converting each density in kg/m^3 to the equivalent value in g/cm^3. The first two have been done for you.

b Use the data to explain why ice floats on water.

....

....

c A cook mixes equal volumes of water and olive oil in a jar. The two liquids separate. Complete the drawing of the jar to show how the liquids will appear. Label them.

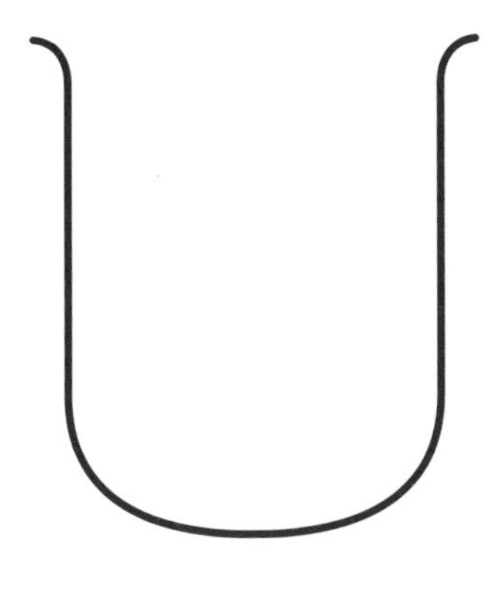

d A student wrote: "This data shows that metals are denser than non-metals." Do you agree or disagree? Explain your answer.

....

....

....

....

e Calculate the mass of a block of gold that measures 20 cm × 15 cm × 10 cm. Give your answer in kg.

f A metalworker finds a block of silvery metal. He weighs it and he measures its volume. Here are his results:

mass of block = 0.270 kg
volume of block = 14.0 cm^3

Calculate the density of the block.

g Suggest what metal this might be. ....

Exercise 1.3 Testing your body clock

How good would your pulse be as a means of measuring time intervals? Galileo used the regular pulse of his heart as a means of measuring intervals of time until he noticed that a swinging pendulum was more reliable.

In this exercise, you need to be able to measure the pulse in your wrist. Place two fingers of one hand gently on the inside of the opposite wrist. Press gently at different points until you find the pulse. (Alternatively, press two fingers gently under the jawbone on either side of your neck.)
You will also need a clock or watch that will allow you to measure intervals of time in seconds.

a Start by timing 10 pulses. (Remember to start counting from zero: 0, 1, 2, 3, . . . , 9, 10.) Repeat this several times and record your results below.

b Comment on your results. How much do they vary? Is the problem that it is difficult to time them, or is your heart rate varying?

....

....

....

....

....

c Use your results to calculate the average time for one pulse.

d Repeat the above, but this time count 50 pulses. Record your results below. Calculate the average time for one pulse.

e Now investigate how your pulse changes if you take some gentle exercise – by walking briskly, or by walking up and down stairs, for example. In the space below:

- Briefly describe your exercise.
- Give the measurements of pulse rate that you have made.
- Comment on whether you agree with Galileo that a pendulum is a better time-measuring instrument than your pulse.

....

....

Describing motion

Definitions to learn

speed the distance travelled by an object in unit time
acceleration the rate of change of an object's velocity

Equations to learn and use

$$\text{speed} = \frac{\text{distance}}{\text{time}}$$

speed = gradient of distance against time graph

distance = area under speed against time graph

E

$$\text{acceleration} = \frac{\text{change in speed}}{\text{time taken}}$$

acceleration = gradient of speed against time graph

Exercise 2.1 Speed calculations

Use the equation for speed to solve some numerical problems.

a The table below shows the time taken for each of three cars to travel 100 m. Circle the name of the fastest car.

Car	Time taken / s	Speed / m/s
red car	4.2	
green car	3.8	
yellow car	4.7	

b Complete the table by calculating the speed of each car. Give your answers in m/s and to one decimal place.

c A jet aircraft travels 1200 km in 1 h 20 min.

How many metres does it travel? ....

For how many minutes does it travel? ....

And for how many seconds? ....

d Calculate the average speed of the jet in part **c** during its flight. Give your answer in m/s.

e A stone falls 20 m in 2.0 s. Calculate its average speed as it falls.

f The stone falls a further 25 m in the next 1.0 s of its fall. Calculate the stone's average speed during the 3 s of its fall.

g Explain why we can only calculate the stone's **average** speed during its fall.

....

....

Exercise 2.2 Distance against time graphs

Draw and interpret some distance against time graphs. You can calculate the speed of an object from the gradient (slope) of the graph.

a The diagrams A–D show distance against time graphs for four moving objects.

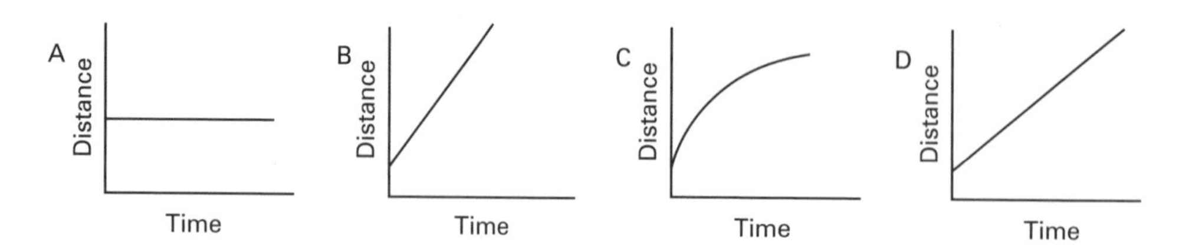

Complete the table below by indicating (in the second column) the graph or graphs that represent the motion described in the first column.

Description of motion	Graph(s)
moving at a steady speed	
stationary (not moving)	
moving fastest	
changing speed	

b The table below shows the distance travelled by a runner during a 100 m race. Use the data to draw a distance against time graph on the graph paper on the next page.

Distance / m	0	10.0	25.0	45.0	65.0	85.0	105.0
Time / s	0.0	2.0	4.0	6.0	8.0	10.0	12.0

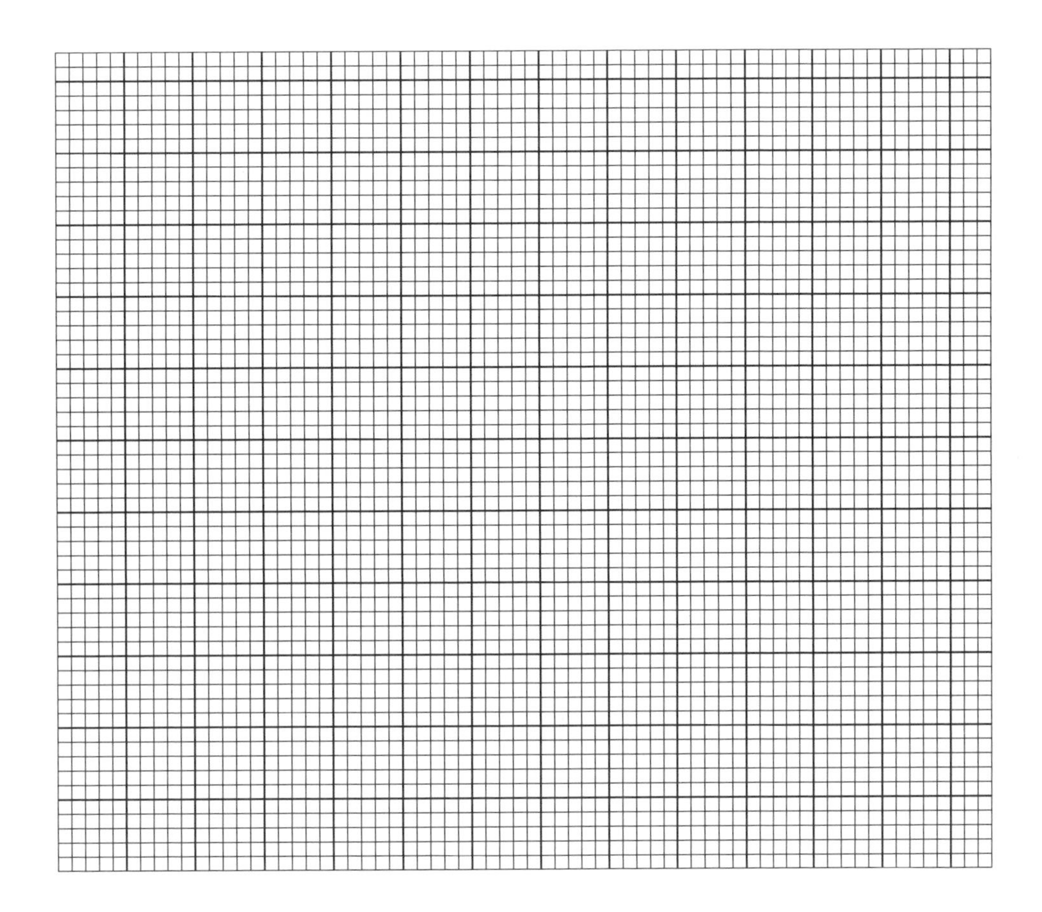

c From your graph, answer the questions that follow:

How far did the runner travel in the first 9.0 s? ....

How long did the runner take to run the first 50.0 m? ....

How long did the runner take to complete the 100 m? ....

d Use the gradient of the graph to determine the runner's average speed between 4.0 s and 10.0 s. On the graph, show the triangle that you use.

e On the graph axes on the next page, sketch a distance against time graph for the car whose journey is described here:

- The car set off at a slow, steady speed for 20 s.
- Then it moved for 40 s at a faster speed.
- Then it stopped at traffic lights for 20 s before setting off again at a slow, steady speed.

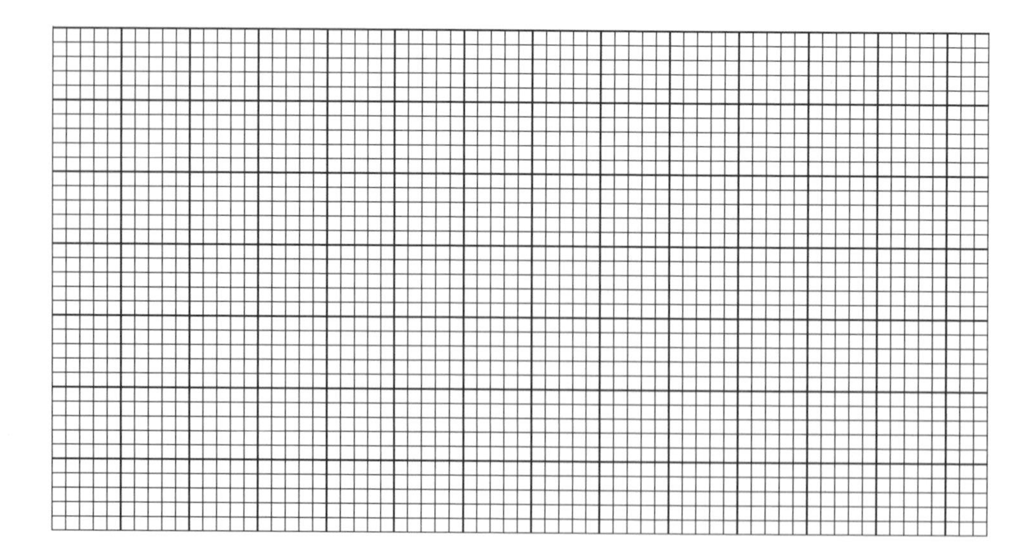

f The graph on the right represents the motion of a bus for part of a journey. On the graph, mark the section of the journey where the bus was moving faster.

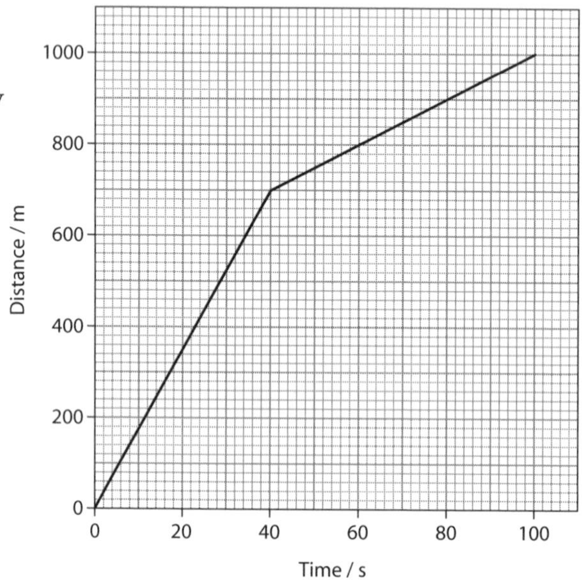

g From the graph in part **f**, calculate the following:

- the speed of the bus when it was moving faster

- the average speed of the bus.

E

Exercise 2.3 Speed against time graphs

Draw and interpret some speed against time graphs. You can calculate the acceleration of an object from the gradient (slope) of the graph. You can calculate the distance travelled from the area under the graph.

a The diagrams show speed against time graphs for four moving objects.

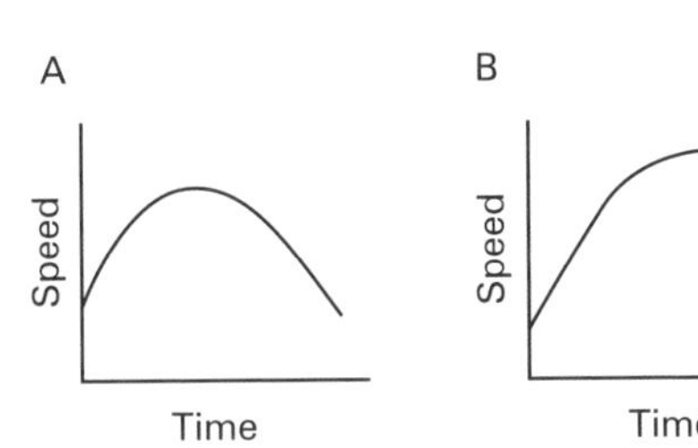

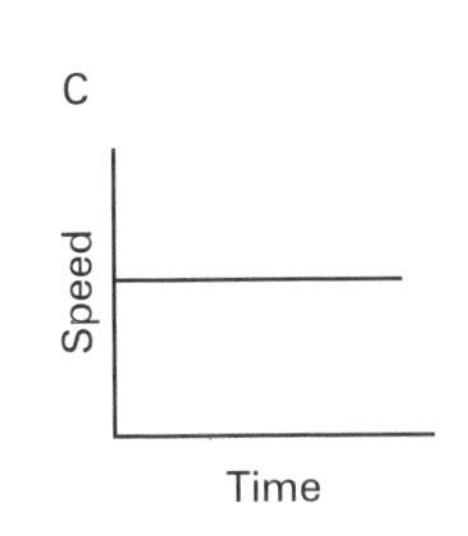

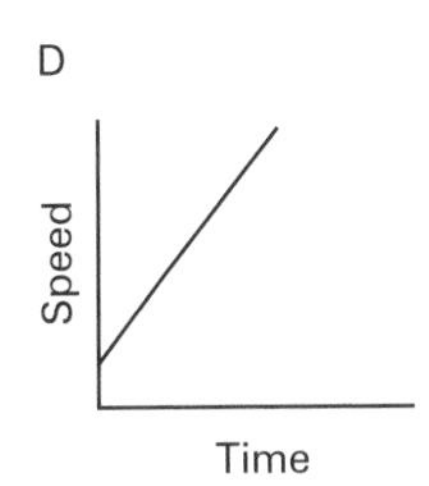

Complete the table by indicating (in the second column) the graph or graphs that represent the motion described in the first column.

Description of motion	Graph(s)
moving at a steady speed	
speeding up, then slowing down	
moving with constant acceleration	
accelerating to a steady speed	

b The graph on the right represents the motion of a car that accelerates from rest and then travels at a steady speed.

From the graph, determine the acceleration of the car in the first part of its journey.

E

c On the graph, shade the area that represents the distance travelled by the car while accelerating. Label this area **A**.

Shade the area that represents the distance travelled by the car at a steady speed. Label this area **B**.

d Calculate each of these distances and the total distance travelled by the car.
[Note: area of a triangle = $\frac{1}{2}$ × base × height]

e On the graph paper below, sketch a speed against time graph for the car whose journey is described here:

- The car set off at a slow, steady speed for 20 s.
- Then, during a time of 10 s, it accelerated to a faster speed.
- It travelled at this steady speed for 20 s.
- Then it rapidly decelerated and came to a halt within 10 s.

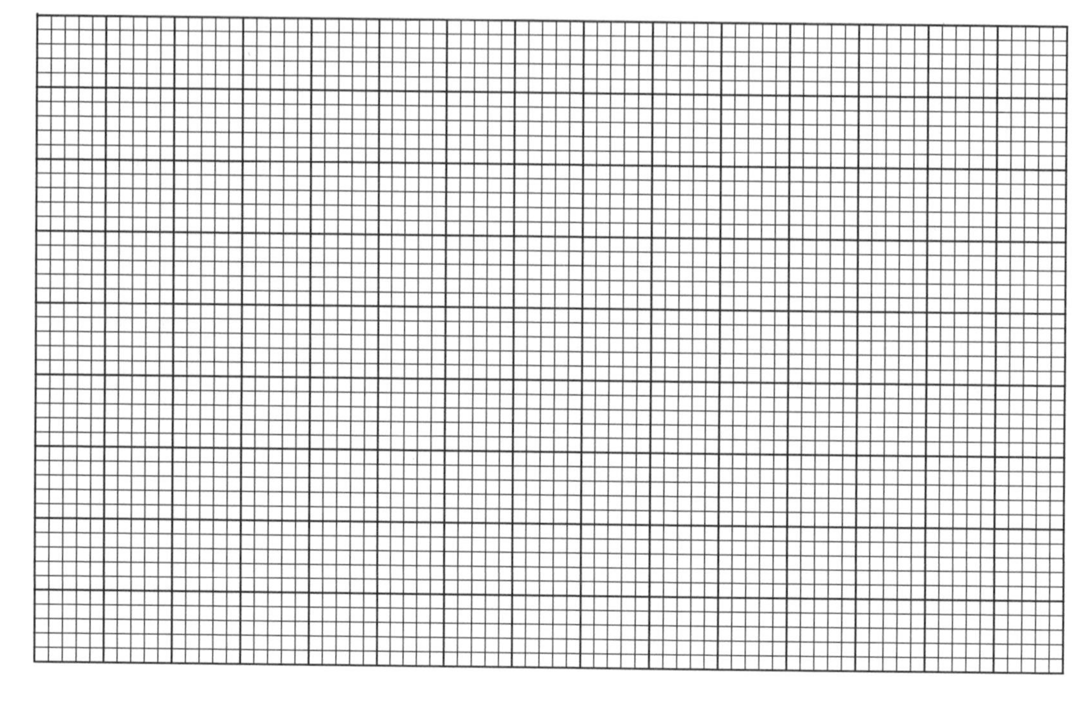

Forces and motion

Definitions to learn

force	an interaction between two bodies
resultant force	the single force that has the same effect on a body as two or more forces
mass	the property of an object that causes it to resist changes in its motion
weight	the downward force of gravity that acts on an object because of its mass
vector quantity	a quantity that has both magnitude and direction
scalar quantity	a quantity that has only magnitude

E

An equation to learn and use

E

force = mass × acceleration $F = ma$

Exercise 3.1 Identifying forces

Forces are invisible (although we can often see their effects). Being able to identify forces is an important skill for physicists.

The pictures show some bodies. Your task is to add at least one force arrow to each body, showing a force acting on it. (Two force arrows are already shown.)
Each force arrow should be labelled to indicate the following:

- the type of force (e.g. contact, drag/air resistance, weight/gravitational, push/pull, friction, magnetic)
- the body causing the force
- the body acted on by the force.

For example: the gravitational force of the Earth on the apple

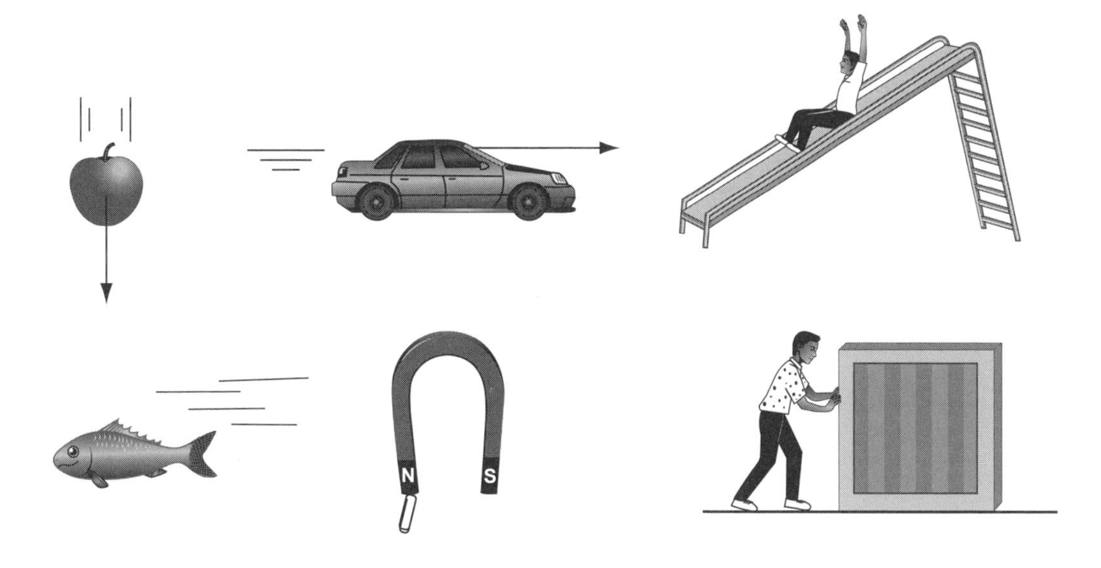

Exercise 3.2 The effects of forces

A force can change how a body moves, or it may change its shape.

a Each diagram below shows a body with a single force acting on it. For each, say what effect the force will have.

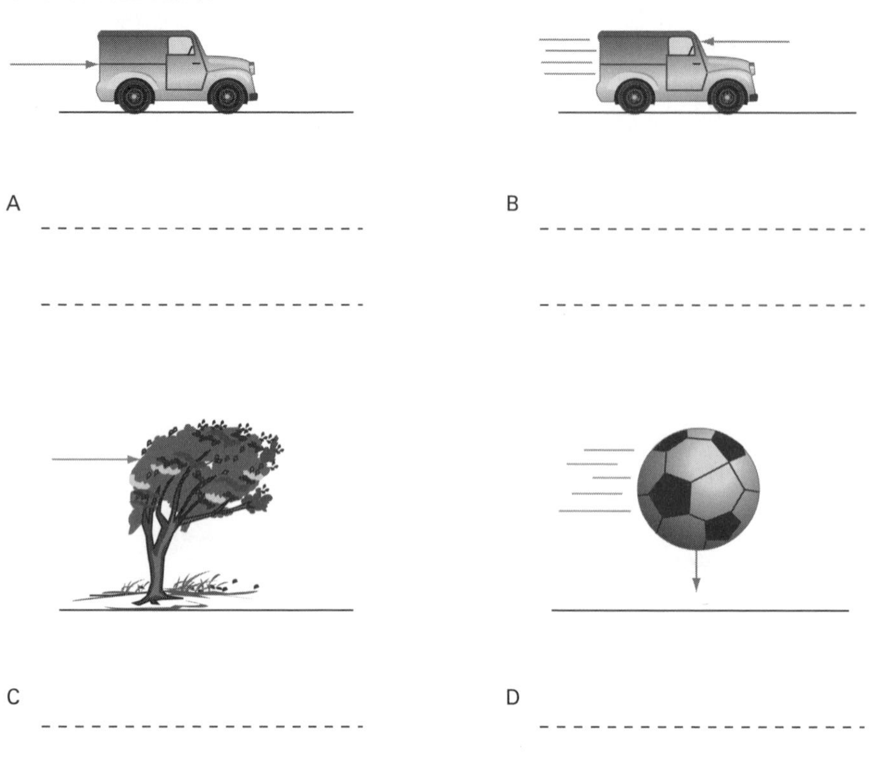

b A boy slides down a sloping ramp. In the space below, draw a diagram of the boy on the ramp and add a labelled arrow to show the force of friction that acts on him.

c What effect will the force have on the boy's movement?

....

....

Exercise 3.3 Falling

What is the pattern of motion of a falling object? How do the forces of gravity and friction affect a falling body?

Galileo is said to have dropped two objects of different masses from the top of the Leaning Tower of Pisa in Italy. The diagram shows the position of the smaller object at equal intervals of time as it fell.

a The spacings between the dots gradually increase. What does this tell you about the speed of the falling object?

....

....

b Add dots to the diagram to show the pattern you would expect to find for the object with greater mass (at the same intervals of time).

c What can you say about the accelerations of the two objects?

....

....

d Galileo's young assistant would probably have enjoyed attaching a parachute to a stone and dropping it from the tower. After a short time, the stone would fall at a steady speed. Add some more dots to the diagram above to show the pattern you would expect to see for this.

e The graph below shows how the stone's speed would change as it fell. On the right are two drawings of the stone. These correspond to points **A** and **B** on the graph.

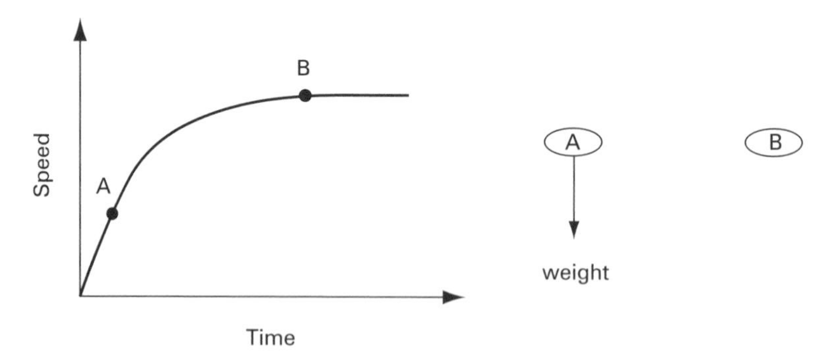

Diagram **A** shows the stone's weight. Add a second force arrow to this diagram to show the force of air resistance acting on the stone.
Add two force arrows to diagram **B** to show the forces acting on the stone at this point in its fall.

Turning effects of forces

Definitions to learn

moment of a force	the turning effect of a force about a point
equilibrium	when no net force and no net moment act on a body
centre of mass	the point at which the mass of an object can be considered to be concentrated

Equations to learn and use

E

moment of a force = force × perpendicular distance from pivot to force
For a body in equilibrium:
total clockwise moment = total anticlockwise moment

Exercise 4.1 Turning effect of a force

When a force acts on a body that is pivoted, it can have a turning effect. The body may start to rotate.

a The diagram on the right shows a wheelbarrow with a heavy load of soil. Add an arrow to show how you could lift the left-hand end of the barrow with the smallest force possible. Remember to indicate clearly the direction of the force.

b The diagram below shows a beam balanced on a pivot. Add arrows to show the following forces:

- A 100 N force pressing downwards on the beam that will have the greatest possible clockwise turning effect. Label this 'Force A'.
- A 200 N force pressing downwards on the beam that will have an anticlockwise turning effect equal in size to the turning effect of Force A. Label this 'Force B'.

c If a body is in equilibrium, what can you say about the resultant force on the body?

....

d If a body is in equilibrium, what can you say about the resultant turning effect on the body?

....

Exercise 4.2 Calculating moments

Calculate some moments. Remember that it is important to note whether a moment acts clockwise or anticlockwise.

a In the diagram on the right, all forces are of equal size.

Which force has the greatest moment about point **A**?

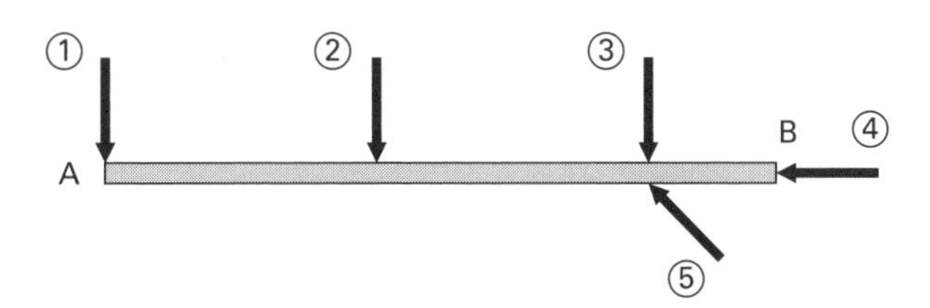

....

b Which force has no moment about point **B**?

....

c Look at the diagram.

Which distance should be used in calculating the moment of force *F* about point **X**?

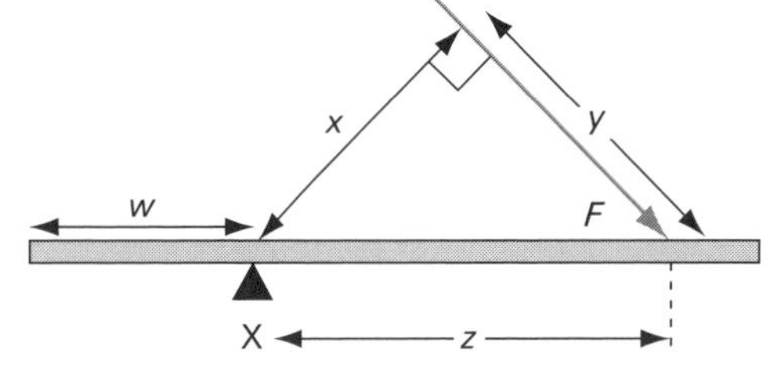

....

E

d Explain your choice of answer for part **c**.

....

....

....

e Calculate the moment about the pivot of each force in the diagram on the right.

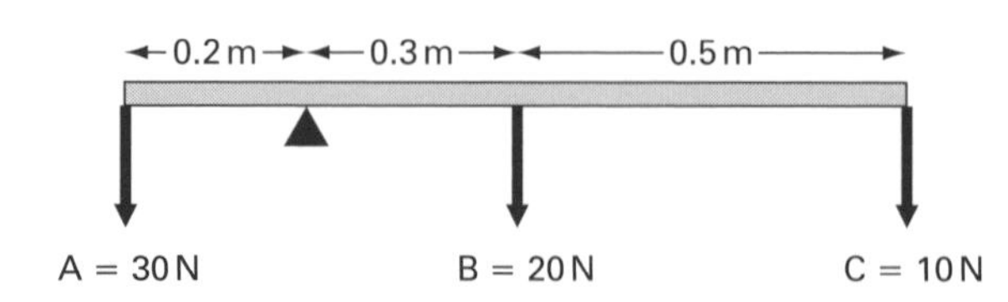

Force	Moment	Clockwise or anticlockwise?
A		
B		
C		

f Which force must be removed if the beam is to be balanced?

....

g In the diagram on the right, the beam is balanced (in equilibrium). Calculate the size of force *F*.

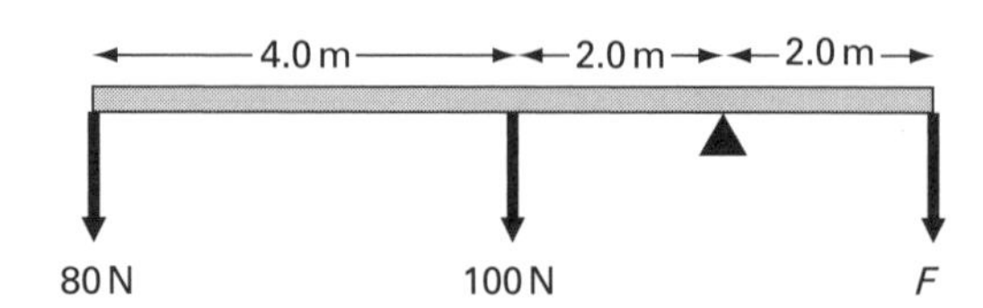

5 Forces and matter

Definitions to learn

extension the increase in length of a spring when a load is attached
limit of proportionality the point beyond which the extension of an object is no longer proportional to the load producing it
load a force that causes a spring to extend
pressure the force acting per unit area at right angles to a surface

Equations to learn and use

E

Hooke's law: force = stiffness × extension

$$\text{pressure} = \frac{\text{force}}{\text{area}} \qquad p = \frac{F}{A}$$

pressure in a fluid = $h\rho g$

Exercise 5.1 Stretching a spring

Robert Hooke discovered his law of springs by attaching weights and measuring the extension of the spring.

a Use mathematical symbols to turn the following into an equation. There are **two** different ways to do it. Can you find both?

stretched length original length extension

stretched length original length extension

b A student carried out an experiment to stretch a spring.

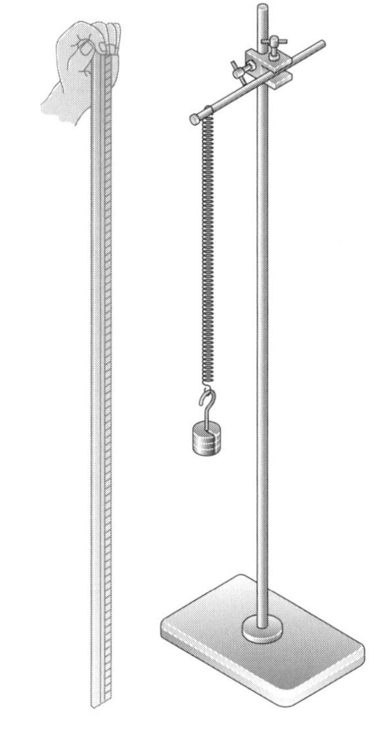

The table shows the results. Complete the third column of the table.

Load / N	Length / cm	Extension / mm
0	25.0	
1.0	25.4	
2.0	25.8	
3.0	26.2	
4.0	26.6	
5.0	27.0	
6.0	27.4	
7.0	27.8	
8.0	28.5	
9.0	29.2	
10.0	29.9	

c From the data in the table, estimate the force needed to produce an extension of 1.0 cm.

....

d On the graph paper, draw a load against extension graph for the spring.

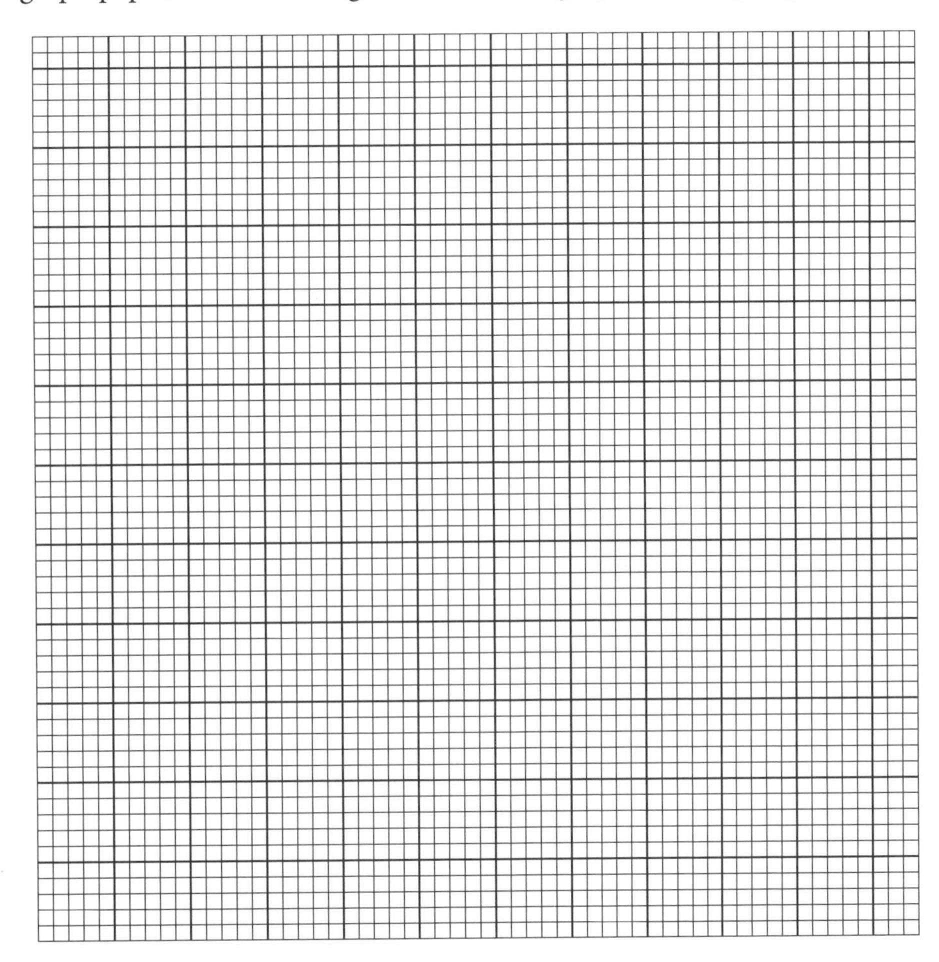

e From your graph, what is the load at the limit of proportionality?

....

Exercise 5.2 Pressure

Has an elephant ever stood on your foot? Pressure explains why it might not hurt quite as much as you think!

a The equation $p = \frac{F}{A}$ is used to calculate pressure. Complete the table to show the name and SI unit of each quantity (name and symbol).

Quantity	Symbol	SI unit
	p	
	F	
	A	

E

b Rearrange the equation in part **a** to make F and A the subjects:

$F =$ $A =$

c It is dangerous to stand on the icy surface of a frozen pond or lake. Explain why it is even more dangerous to stand on one foot than on both feet.

....

....

....

....

d Describe how you could move across the ice in such a way as to minimise the danger of falling through.

....

....

....

e Calculate the pressure when a force of 200 N presses on an area of $0.40\,m^2$.

f The pressure inside a car tyre is 250 kPa (250 000 Pa). Calculate the total force exerted on the inner surface of the tyre if its surface area is $0.64\,m^2$.

g Calculate the pressure at the bottom of an oil storage tank of depth 2.50 m. The oil has a density of $980\,kg/m^3$ and $g = 10\,m/s^2$.

E

h Use the following data to estimate the height of the Earth's atmosphere:

atmospheric pressure = 100 kPa
density of air = 1.29 kg/m^3.

i Explain why this can only be an estimate.

....

....

....

6 Energy transformations and energy transfers

Definitions to learn

kinetic energy	the energy of a moving object
chemical energy	energy stored in chemical substances and which can be released in a chemical reaction
thermal (heat) energy	energy being transferred from a hotter place to a colder place because of the temperature difference between them
electrical energy	energy transferred by an electric current
light energy	energy emitted in the form of visible radiation
strain energy	energy of an object due to its having been stretched or compressed

nuclear energy	energy stored in the nucleus of an atom
internal energy	the energy of an object; the total kinetic and potential energies of its particles
efficiency	the fraction of energy that is converted into a useful form

Equations to learn and use

$$\text{efficiency} = \frac{\text{useful energy output}}{\text{energy input}} \times 100\%$$

gravitational potential energy = weight × height   g.p.e. = mgh

kinetic energy = $\frac{1}{2}$ × mass × speed^2 k.e. = $\frac{1}{2}mv^2$

The principle of conservation of energy: in any energy conversion, the total amount of energy before and after the conversion is constant.

E

Exercise 6.1 Energy efficiency

In many energy transformations, some of the energy is wasted – it ends up in a different form from the one we want. Many energy transformations waste energy as thermal (heat) energy.

a A washing machine has a motor that turns the drum. In a particular washing machine, the motor is supplied with 1200 J of energy each second. Of this, 900 J of energy is used to turn the drum. The rest is wasted as thermal energy. Calculate the amount of energy wasted as heat each second.

b Calculate the efficiency of the motor. Give your answer as a percentage.

c Explain why we say that energy is 'wasted' as thermal energy.

....

....

Here is some information about two power stations:

- A gas-fired station is supplied with 1000 MJ of energy each second and produces 450 MJ of electrical energy.
- A coal-fired power station is supplied with 600 MJ of energy each second and produces 150 MJ of electrical energy.

E

d Calculate the efficiency of each power station.

e Which power station in part **d** is more efficient? 

f A Sankey diagram can be used to represent energy changes. The diagram below shows the energy changes in a light bulb each second.

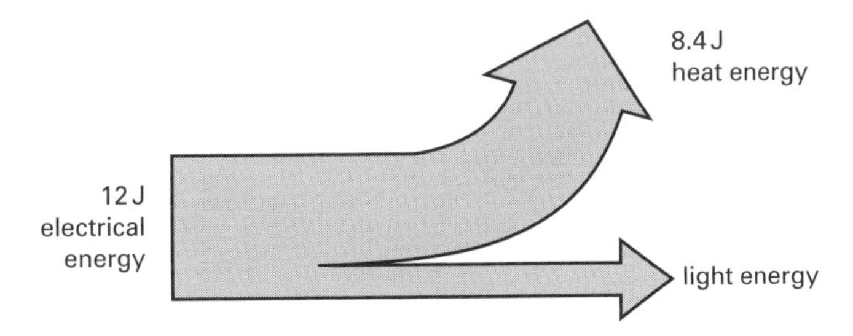

On the diagram, indicate the amount of light energy produced each second.

g Calculate the efficiency of the bulb.

h In the space below, draw a Sankey diagram for the washing machine described in part **a** above.

E

Exercise 6.2 Energy calculations

Because we can calculate quantities of energy, we can make predictions about the outcomes of energy changes. You need to be able to calculate kinetic energy and gravitational potential energy.

a Calculate the kinetic energy of a car of mass 600 kg travelling at 25 m/s.

b The car in part **a** slows down to a speed of 12 m/s. By how much has its kinetic energy decreased?

c A walker carries a 20 kg pack on his back. He climbs to the top of a mountain 2500 m high. Calculate the gain in gravitational potential energy of the pack. (Acceleration due to gravity g = 10 m/s^2.)

Here is an example of how energy calculations can be used to solve problems.

A girl throws a ball upwards. The ball has a mass of 0.20 kg and it leaves her hand with a speed of 8 m/s.

d How high will it rise?

Step 1: Calculate the k.e. of the ball as it leaves the girl's hand.

E

Step 2: When the ball reaches its highest point, it no longer has any k.e. – all of its energy has been transformed to g.p.e. So now we can write:

g.p.e. at highest point = k.e. at lowest point $\quad mgh = \text{k.e.}$

and rearranging gives $\quad h = \dfrac{\text{k.e}}{mg}$

Use this equation to calculate the height to which the ball rises.

e In a game, a toy car slides down a slope. If the top of the slope is 2.0 m higher than the foot of the slope, how fast will the car be moving when it reaches the foot? (Assume that all of its g.p.e. is transformed to k.e.)

7 Energy resources

Definitions to learn

biomass fuel a material, recently living, used as a fuel
fossil fuel a material, formed from long-dead material, used as a fuel
renewable energy resource that, when used, will be replenished naturally
non-renewable energy resource that, once used, is gone forever
geothermal energy the energy stored in hot rocks underground
photocell, solar cell an electrical device that transfers the energy of sunlight directly to electricity, by producing a voltage when light falls on it

nuclear fission	the process by which energy is released by the splitting of a large heavy nucleus into two or more lighter nuclei
nuclear fusion	the process by which energy is released by the joining together of two small light nuclei to form a new heavier nucleus

Exercise 7.1 Renewables and non-renewables

Most of the energy we use comes from non-renewable sources. If we used only renewable resources, our way of life would be more sustainable.

a Complete the table below as follows:

- In the second column, write the name of the energy resource described in the first column.
- In the third column, indicate whether the resource is renewable or non-renewable.

Description	Energy resource	Renewable / non-renewable
wood		
natural gas		
coal		
splitting of uranium nuclei		
hydrogen nuclei combine to release energy		
sunlight captured to make electricity		
hot rocks underground used to heat water		
moving air turns a turbine		
water running downhill turns a turbine		

b In the space below, draw a diagram (with added notes) to explain why hydro-power can be described as renewable.

Work and power

Definitions to learn

doing work transferring energy by means of a force
energy the capacity to do work
joule (J) the SI unit of work or energy
power the rate at which energy is transferred or work is done
work done the amount of energy transferred when one body exerts a force on another; the energy transferred by a force when it moves

Equations to learn and use

work done = energy transferred $\Delta W = \Delta E$

work done = force × distance moved by the force $\Delta W = F \times d$

$$\text{power} = \frac{\text{work done}}{\text{time taken}}$$

$$\text{power} = \frac{\text{energy transferred}}{\text{time taken}}$$

Exercise 8.1 Forces doing work, transferring energy

When a force moves, it does work. It transfers energy to the object it is acting on. Use these ideas to answer some questions.

a Complete these sentences:

An apple falls from a tree. The force acting on the apple to make it fall is

.... As it falls, its speed .... This shows

that its .... energy is increasing. If its energy increases by 2.0 J,

the work done on it is ....

b The girl in the picture is raising a heavy load.

How can you tell that the load's energy is increasing?

....

....

c Explain where this energy in part **b** comes from.

....

....

d Explain how the energy in part **b** is transferred to the load.

....

....

e In the picture, the 20 N force does more work than the 10 N force.

10 N

20 N

State **two** ways that you can tell this.

....

....

....

....

E

Exercise 8.2 Calculating work done

Practise calculating the work done by a force when it moves.

a A boy pushes a heavy box along the ground. His pushing force is 75 N. He pushes it for a distance of 4.0 m. Calculate the work done by the boy in pushing the box.

b A girl lifts a heavy box above her head to place it on a shelf.

Her lifting force is 120 N.
She lifts the box to a height of 1.6 m.

Calculate the work done by the girl in lifting the box.

E

c The girl decides it would be easier to push the box up a sloping ramp. Her pushing force is 80 N. The length of the ramp is 3.0 m. Calculate the work done by the girl.

d Give the reason why more work was done pushing the box up the ramp than lifting it straight up.

....

....

Exercise 8.3 Power

Power is the rate at which a force does work, or the rate at which energy is transferred. Practise calculations involving power.

a A light bulb is labelled with its power rating: 60 W. How many joules of energy does it transfer in 1 s? ....

b How many joules of energy does the light bulb in part **a** transfer in 1 minute?

....

c Why would it be incorrect to say that the light bulb in part **a** supplies 60 J of light energy each second?

....

....

....

E

d A growing person needs a diet that supplies about 10 MJ of energy per day. Calculate the amount of energy supplied by such a diet each second, and hence the person's average power. (Give your answer to the nearest 10 W.)

A motor car is travelling at a steady speed of 30 m/s. The engine provides the force needed to oppose the force of air resistance, 1600 N.

e In the space below, draw a diagram to show the four forces that act on the car.

f Calculate the work done by the car in each second against the force of air resistance.

g What power is supplied by the car's engine? ....

Block 2
Thermal physics

9 The kinetic model of matter

Definitions to learn

boiling point	the temperature at which a liquid changes to a gas (at constant pressure)
melting point	the temperature at which a solid melts to become a liquid
evaporation	when a liquid changes to a gas at a temperature below its boiling point
Brownian motion	the motion of small particles suspended in a liquid or gas, caused by molecular bombardment
kinetic model of matter	a model in which matter consists of molecules in motion

An equation to learn and use

E

Boyle's law: $p_1V_1 = p_2V_2$

Exercise 9.1 Changes of state

Ice, water, steam – these are all the same substance in different states. How well do you know the three different states of matter?

a Which states of matter are being described here? Complete the table.

Description	State or states
occupies a fixed volume	
evaporates to become a gas	
takes the shape of its container	
may become a liquid when its temperature changes	

b Label each arrow in the diagram below to show the name of the change of state.

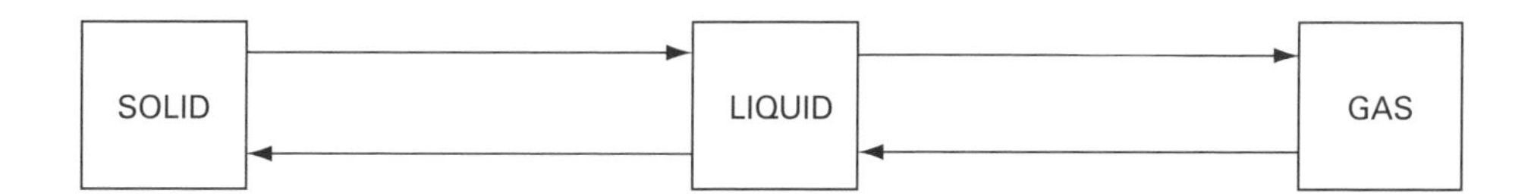

Salt solution freezes at a temperature a few degrees below the freezing point of pure water. You have to investigate how the temperature of some salt solution changes as it is cooled from +20 °C to −20 °C. You are provided with an electronic thermometer and a freezer, which is set to give a temperature of −20 °C.

c Describe in words with a diagram how you would set about this task.

....

....

....

....

d On the axes below, sketch the shape of the temperature against time graph you would expect to obtain in this investigation. Indicate how you would use the graph to deduce the freezing point of the salt solution.

Exercise 9.2 Brownian motion

Robert Brown observed tiny fragments of pollen grains moving around in water. He thought they might be alive, but then he saw the same thing with coal dust. He never fully understood why they moved.

The diagram shows equipment for observing Brownian motion.

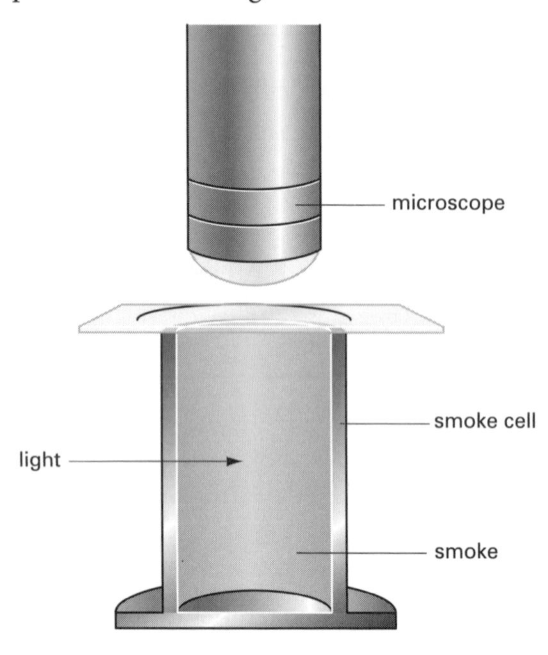

a On the diagram, show how light coming from the left reaches the observer looking down the microscope.

b Why must a microscope be used?

....

....

c Describe briefly what the observer sees.

....

....

....

d Why can we not see molecules of air in the smoke cell?

....

....

e Write a brief explanation of the observations using ideas from the kinetic model of matter.

....

....

....

....

....

....

E

Exercise 9.3 Boyle's law

Robert Boyle investigated how the pressure of a gas depends on its volume.

a Boyle's law can be represented by the equation: pV = constant
Complete the table to show the meanings of the symbols in this equation and their units.

Symbol	Name of quantity	SI unit (name and symbol)
p		
V		

b What **two** quantities must remain constant if Boyle's law is to be obeyed?

....

....

c A cylinder contains 400 cm^3 of air at a pressure of 2.0×10^5 Pa. The gas is compressed to a volume of 160 cm^3. Calculate the pressure when the gas has returned to its original temperature.

E

d Compressing a gas generally increases its temperature. If the pressure of the air in part c had been measured before it had returned to its original temperature, would it have been greater or less than your calculated value? Explain your answer.

....

....

....

....

In an experiment to investigate Boyle's law, a student obtained the values of pressure and volume shown in the table below. Unfortunately, although she planned to record five values, she only recorded three.

Pressure of gas / kPa	Volume of gas / cm^3	Pressure × volume
100	88	
120		
140	63	
160		
180	50	

e Plot these points on the graph axes below and draw a suitable curve through them.

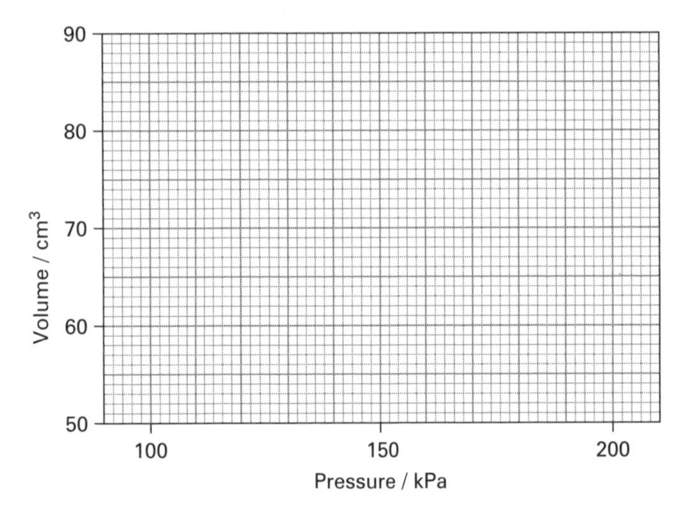

Use your graph to find the volumes for 120 kPa and 160 kPa pressure.

E

f Use your graph to find the missing volumes. Show how you did this on the graph.

g Add the missing volumes to the table.

h Calculate the five values for pressure × volume and add them to the third column.

Thermal properties of matter

Definitions to learn

temperature	a measure of how hot or cold something is
internal energy	the energy of an object; the total kinetic and potential energies of its particles
thermistor	a resistor whose resistance changes a lot over a small temperature range
thermocouple	an electrical device made of two different metals, used as an electrical thermometer
thermal equilibrium	describes the state of two objects (or an object and its surroundings) that are at the same temperature so that there is no heat flow between them
thermal expansion	the expansion of a material when its temperature rises
specific heat capacity (s.h.c.)	a measure of how much energy is required to raise the temperature of 1 kg of a material by 1 °C
specific latent heat	the energy required to melt or boil 1 kg of a substance

Equations to learn and use

E

energy supplied = mass × specific heat capacity × increase in temperature

$$\text{specific latent heat} = \frac{\text{energy supplied}}{\text{mass}}$$

Exercise 10.1 Calibrating a thermometer

All scientific instruments need to be calibrated if they are to provide reliable measurements.

A student has an uncalibrated alcohol-in-glass thermometer. She places it in melting ice and then in boiling water. She measures the length of the alcohol column each time. The table on the next page shows her results.

Condition	Temperature / °C	Length of alcohol column / cm
melting ice		12.0
boiling water		26.8

a Complete the table by filling in the values of the temperatures.

b Explain what it means to say that the thermometer was 'uncalibrated'.

....

....

c You can now draw a calibration graph using the grid below, as follows. Mark the two points corresponding to the data in the table. Join them with a straight line. Use your graph to answer the following three questions. (Mark the graph to show your method.)

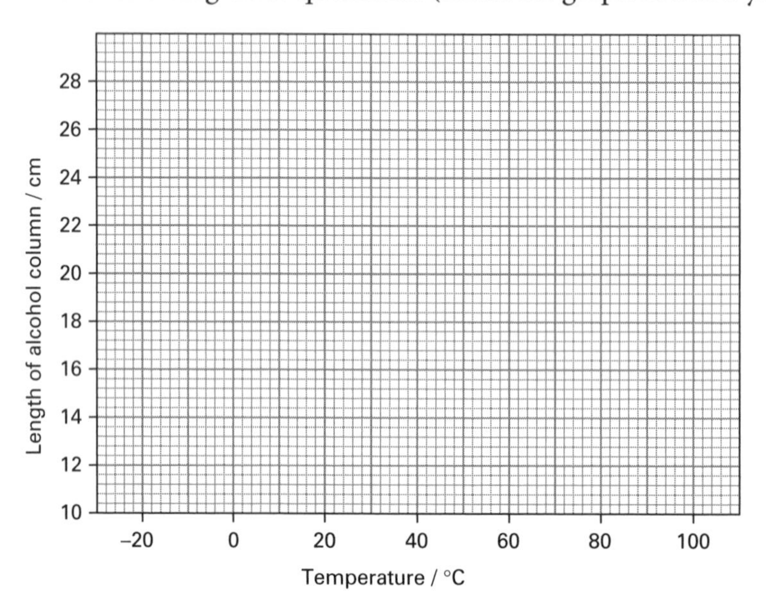

d If the length of the alcohol column is 14.8 cm, what is the temperature?

....

e What will be the length of the alcohol column at a temperature of 60 °C?

....

f The thermometer is placed in a freezer. The length of the alcohol column is 10.0 cm.

What is the temperature inside the freezer? ....

Exercise 10.2 Heat calculations

To make an object hotter, thermal (heat) energy must be supplied. This increases its internal energy. Knowing the specific heat capacity of a material, you can calculate temperature rises and amounts of energy.

The table shows the specific heat capacities of a variety of materials. Use this information to answer the questions that follow.

Type of material	Material	Specific heat capacity / J/(kg °C)
metals	steel	420
	aluminium	910
	copper	385
	gold	300
	lead	130
non-metals	glass	670
	nylon	1700
	polythene	2300
	ice	2100
liquids	water	4200
	sea water	3900
	ethanol	2500
	olive oil	1970
gases	air	1000
	water vapour	2020 (at 100 °C)
	methane	2200

a Which of the metals shown in the table will require the greatest amount of energy to raise the temperature of 100 g from room temperature to 200°C?

....

E

b You have two identical glass beakers containing equal amounts of water and sea water. You heat them using identical electrical heaters, and record their temperatures as they rise. Which temperature will rise more quickly? Explain your answer.

....

....

....

c A 1.0 kg block of steel is heated in an oven to a temperature of 200 °C. It is then dropped into a tank containing 100 kg of water. The experiment is repeated using a 1.0 kg block of aluminium. Which block will cause a bigger rise in the temperature of the water? Explain your answer.

....

....

....

....

d Which of the following statements are true and which are false (based on the information given in the table on the previous page)?

Statement	True / false
All metals have a lower s.h.c. than all non-metals.	
Metals generally have a lower s.h.c. than non-metals.	
The s.h.c. of water decreases when it freezes.	
The s.h.c. of water decreases when it boils.	

e How much energy must be supplied to a 5.0 kg block of copper to increase its temperature from 20 °C to 100 °C?

E

f In an experiment to determine the s.h.c. of lead, a 0.80 kg block of lead is heated using a 60 W electric heater for 5.0 minutes. Calculate the energy supplied by the heater in this time.

g The temperature of the block is found to have increased from 20 °C to 165 °C. Use this information to estimate the s.h.c. of lead.

h The value you obtain should be higher than that given in the table above. Suggest **two** reasons why this might be.

....

....

....

....

i The specific latent heat of fusion of ice has a value of 330 000 J/kg. Give another word that could be used instead of 'fusion' in the sentence above. ....

j Calculate the energy that must be supplied to melt 200 g of ice.

Thermal (heat) energy transfers

Definitions to learn

conduction	the transfer of thermal (heat) energy or electrical energy through a material without the material itself moving
convection	the transfer of thermal (heat) energy through a material by movement of the material itself
radiation	energy spreading out from a source carried by particles or waves
electromagnetic radiation	energy travelling in the form of waves
infrared radiation	electromagnetic radiation whose wavelength is greater than that of visible light; sometimes known as heat radiation

Exercise 11.1 Convection currents

Convection is a mechanism by which thermal (heat) energy can spread around by movement of a gas or liquid. These questions will test your understanding of convection.

a The diagram shows a room with a heater next to one wall, opposite a window. Add to the diagram to show how a convection current will form in the room when the heater is switched on.

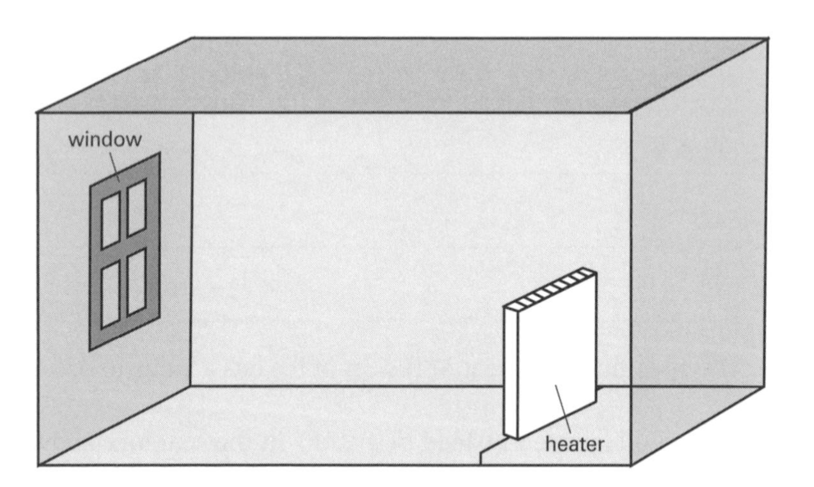

b Explain why it would not be sensible to fix the heater high up on the wall, close to the ceiling.

....

....

....

....

c How do the following quantities change when air is heated? Choose from:

increases   decreases   stays the same

- Temperature ....
- Mass ....
- Density ....
- Separation of molecules ....
- Speed of molecules ....

d Why does the smoke produced by a candle flame rise upwards? Give a detailed explanation.

....

....

....

....

....

....

....

....

....

....

Exercise 11.2 Radiation

Radiation is another mechanism by which thermal (heat) energy can spread around. In this case, it travels in the form of waves, which we call infrared radiation.

a Explain why energy can reach us from the Sun by radiation but not by conduction or convection.

....

....

....

....

b Infrared radiation is just one form of which type of radiation? ....

c Name **one** other form of this type of radiation. ....

d Infrared radiation may be absorbed when it reaches the surface of an object. Describe the surface of an object that is a good absorber of infrared radiation.

....

e What effect does infrared radiation have on an object that absorbs it?

....

....

In cold countries, windows are often fitted with double glazing. This consists of two sheets of glass separated by a gap a few millimetres wide. There is a vacuum in the gap.

f Explain why energy cannot escape from the room by conduction.

....

....

....

g Explain why energy cannot escape from the room by convection.

....

....

....

h Can energy escape by radiation? Explain your answer.

....

....

....

A television remote control uses infrared radiation to send instructions to the TV set. If you point it in the wrong direction, the beam misses the TV set and nothing happens. However, infrared can be reflected by hard, shiny surfaces, such as glass or aluminium.

i In the space below, draw a diagram to show how you could use a remote control, a TV set and a sheet of aluminium to show the reflection of infrared radiation. (You may be able to try this experiment at home. Use a large china plate instead of the metal sheet.)

(Although our eyes cannot see infrared radiation, a digital camera may detect it. Try shining a TV remote control into a camera. Can you see the camera light up when you press the buttons on the remote control?)

Block 3

Physics of waves

Sound

Definitions to learn

amplitude the greatest height of a wave above its undisturbed level
infrasound sound waves whose frequency is so low that they cannot be heard
ultrasound sound waves whose frequency is so high that they cannot be heard
period the time for one complete vibration or the passage of one complete wave
frequency the number of vibrations or waves per second
pitch how high or low a note sounds

E

compression a region of a sound wave where the particles are pushed close together
rarefaction a region of a sound wave where the particles are further apart

An equation to learn and use

E

$$\text{frequency} = \frac{1}{\text{period}} \qquad f = \frac{1}{T}$$

Exercise 12.1 Sound on the move

Sound is a way in which energy can travel from place to place. It can be detected by our ears. Check your basic ideas about sound.

a What one word describes the movement of a source of sound? ....

b Which part of a guitar moves to produce a sound? ....

c What moves when a wind instrument such as a trumpet produces a sound?

....

d What do we call a reflected sound? ....

e Bats and other creatures can find their way around using sound whose pitch is too high for us to hear. What name is given to this form of sound? ....

f Some animals such as elephants can hear notes that are too low for humans to hear.

What name is given to this form of sound? ....

g You can probably hear notes of higher pitch than your teacher. How would you show this in the school laboratory?

....

....

....

....

....

h The speed of sound in air is about 330 m/s. How long will it take sound to travel 1 km in air? (Give your answer in seconds, to one decimal place.)

i In an experiment to measure the speed of sound in glass, a pulse of sound is sent into a glass rod, 14.0 m in length. The reflected sound is detected after 5.6 ms (0.0056 s). Calculate the speed of sound in glass.

The picture shows a method for determining the speed of sound.

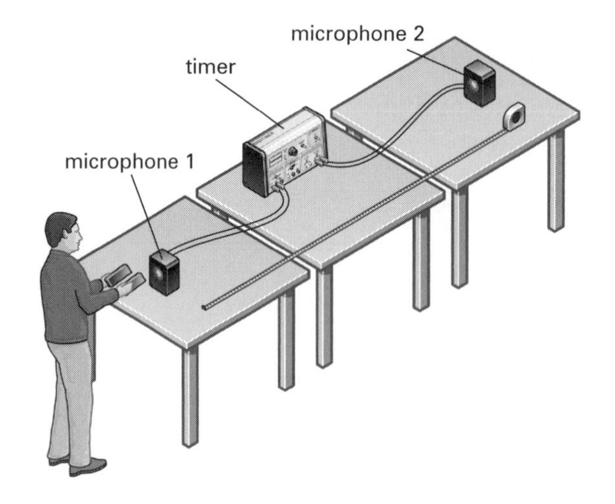

Complete the following four sentences:

j This experiment measures the speed of sound in ....

k To make a sound

....

....

l The microphones detect the sound and the timer shows

....

....

....

m The boy must also measure

....

....

n The formula for calculating the speed of sound from this experiment is:

Exercise 12.2 Sound as a wave

Although we can simply think of sound as energy travelling from place to place, we can understand its properties better if we think of it as a wave.

a Can sound waves travel through a vacuum (empty space)? ....

b What instrument do we use to detect sound waves? ....

c What instrument do we use to display sound waves on a screen? ....

d The diagram shows a trace that represents a sound wave. Add labelled arrows to show the amplitude *A* of the wave and its period *T*.

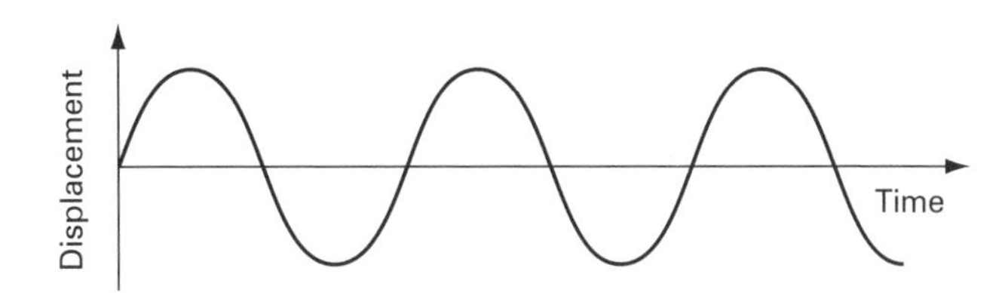

e The diagram shows a trace that represents a sound wave. Add to the diagram to show a second wave that has the same pitch but is louder.

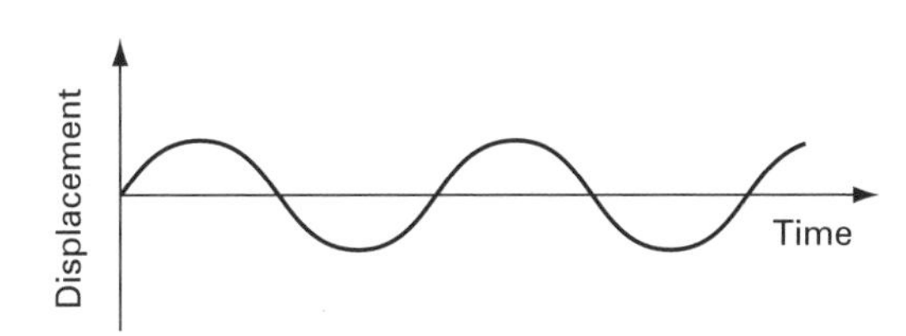

f Two sound waves have the frequencies shown:

sound A 440 Hz sound B 520 Hz

Which sound has the higher pitch? ....

g Calculate the period of sound A in part f.

h A drummer strikes the horizontal surface of a drum so that a sound wave travels upwards from the surface of the drum. Describe how a molecule of the air above the drum will move as the sound wave travels upwards. (It may help to include a simple diagram.)

....

....

....

....

Light

Definitions to learn

reflection	the change in direction of a ray of light when it strikes a surface without passing through it
ray diagram	a diagram showing the paths of typical rays of light
real image	an image that can be formed on a screen
virtual image	an image that cannot be formed on a screen; formed when rays of light appear to be spreading out from a point
refraction	the bending of a ray of light on passing from one material to another
refractive index	the property of a material that determines the extent to which it causes rays of light to be refracted
speed of light	the speed at which light travels (usually in a vacuum: 3.0×10^8 m/s)
principal focus	the point at which rays of light parallel to the axis converge after passing through a converging lens
axis	the line passing through the centre of a lens, perpendicular to its surface
total internal reflection (TIR)	when a ray of light strikes the inner surface of a material and 100% of the light reflects back inside it
critical angle	the minimum angle of incidence at which total internal reflection occurs

Equations to learn and use

Law of reflection: angle of incidence = angle of reflection $i = r$

E

Refractive index: $n = \frac{\text{speed of light in a vacuum}}{\text{speed of light in the material}}$

Snell's law: $n = \frac{\sin i}{\sin r}$

Exercise 13.1 On reflection

Ray diagrams are used to predict where an image will be formed. They can be used when light rays are reflected or refracted.

The incomplete ray diagram below shows an object in front of a plane mirror. Three light rays are shown leaving the object. Follow the instructions to complete the diagram. Then answer the three questions.

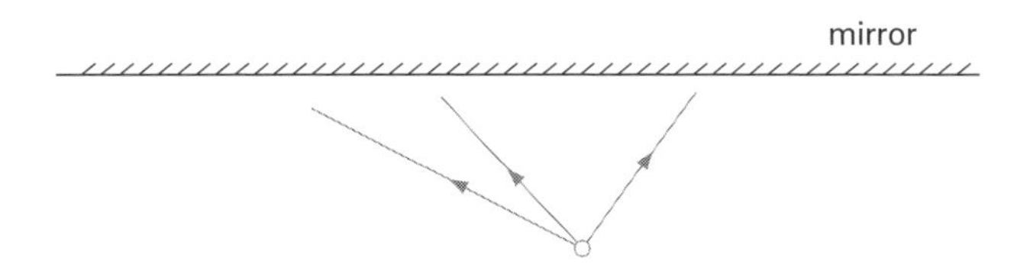

Extend the rays to the mirror.

For each ray, use a ruler and protractor to draw the reflected ray.

Extend the reflected rays to find where they meet behind the mirror.

Mark the position of the image.

Measure the distance of the image from the mirror.

a What is the distance of the image from the mirror? ....

b Is this image real or virtual? ....

c Explain how you know.

....

....

....

Exercise 13.2 Refraction of light

A ray of light is refracted when it passes from one transparent material to another.

The diagram below shows a ray of light travelling from air into glass. Follow the instructions to complete the diagram. Then answer the three questions.

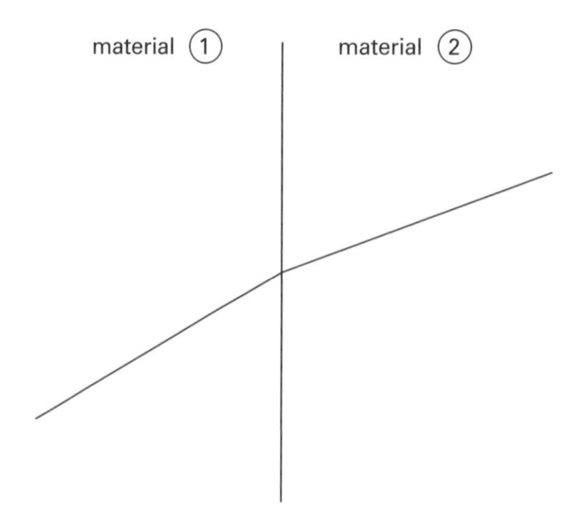

Label the materials 'air' and 'glass'.

Add arrows to the rays to show the direction in which the light is travelling.

Using a ruler, draw the normal to the surface at the point where the ray enters the glass.

Add labels 'incident ray' and 'refracted ray'.

Using a protractor, measure the angle of incidence and the angle of refraction.

a Explain how you know which material is 'air' and which is 'glass'.

....

....

....

b What value did you measure for the angle of incidence? ....

c What value did you measure for the angle of refraction? ....

14 Properties of waves

Definitions to learn

amplitude	the greatest height of a wave above its undisturbed level
frequency	the number of vibrations or waves per second
period	the time for one complete oscillation of a pendulum, one complete vibration or the passage of one complete wave

wavelength	the distance between adjacent crests (or troughs) of a wave
wave speed	the speed at which a wave travels
longitudinal wave	a wave in which the vibration is forward and back, along the direction in which the wave is travelling
transverse wave	a wave in which the vibration is at right angles to the direction in which the wave is travelling
wavefront	a line joining adjacent points on a wave that are all in step with each other
diffraction	when a wave spreads out as it travels through a gap or past the edge of an object

Equations to learn and use

E

wave speed = frequency × wavelength $v = f\lambda$

frequency = $\frac{1}{\text{period}}$ $f = \frac{1}{T}$

Exercise 14.1 Describing waves

A wave transfers energy from place to place without any matter being transferred. There are many different types of wave – sound, light, water – but they all have certain things in common. Do you understand the physicists' model of waves?

The diagram below represents a wave. The *y*-axis shows how much the wave is disturbed from its undisturbed level.

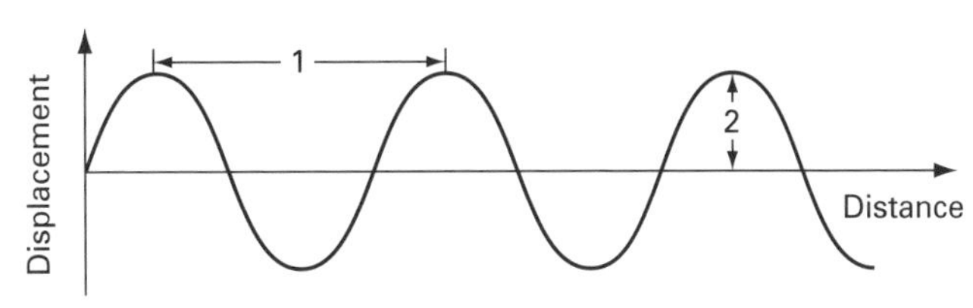

a What does the *x*-axis show? ....

b What quantity does the horizontal arrow 1 indicate? ....

c What symbol is used for this quantity? ....

d What units is it measured in? ....

e What quantity does the vertical arrow 2 indicate? ....

f What symbol is used for this quantity? ....

The diagram below represents a wave. This graph has the time t on the x-axis.

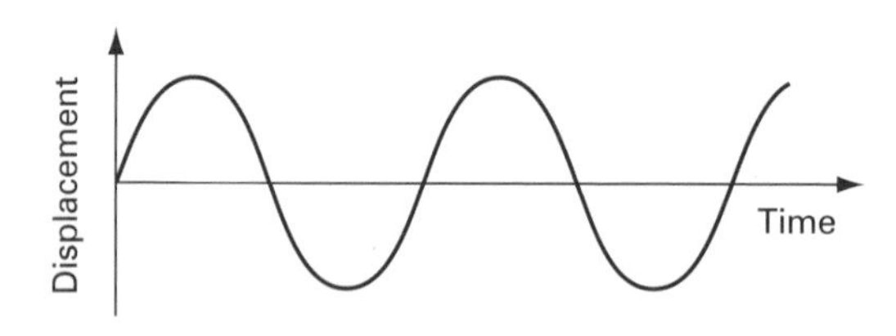

g On the graph, label a crest and a trough.

h Mark a time interval that represents the period T of the wave. Label this T.

i The period of the wave is 0.002 s. Calculate its frequency f. Be sure to give the correct unit.

j Waves can be described as transverse or longitudinal. In which type of wave are the vibrations at right angles to the direction in which the wave is travelling?

....

k Which type of wave is a sound wave: transverse or longitudinal? ....

l Which type of wave is a light wave: transverse or longitudinal? ....

m You have a long spring stretched out in front of you on a long table. Another student holds the far end so that it cannot move. How should you move your end of the spring to produce a transverse wave?

....

....

n How should you move your end of the spring to produce a longitudinal wave?

....

....

E

Exercise 14.2 The speed of waves

The speed of a wave is the speed of a wave crest (or trough) as the wave travels along. Wave speed is related to frequency and wavelength by the equation $v = f\lambda$. These questions will test your understanding of this equation.

a Complete the table to show the quantities related by the equation $v = f\lambda$ and their units.

Symbol	Quantity	Unit (name and symbol)
v		
f		
λ		

b A particular sound wave has a frequency of 100 Hz. How many waves pass a point in 1 s?

....

c If each wave in part **b** has a wavelength of 3.3 m, what is the total length of the waves

that pass a point in 1 s? ....

d So, what is the speed of the sound wave in parts **b** and **c**? ....

Seismic waves are caused by earthquakes. They travel out from the affected area and can be detected around the world. They have low frequencies (mostly too low to hear) and travel at the speed of sound.

e A particular seismic wave is travelling through granite with a speed of 5000 m/s. Its frequency is 8.0 Hz. Calculate its wavelength.

f If the wave is detected 12.5 minutes after the earthquake, estimate the distance from the detector to the site of the quake.

g Explain why your answer to part **f** can only be an estimate.

....

....

....

....

h Light travels at a speed of 3×10^{8} m/s. Red light has a wavelength of 7.0×10^{-7} nm. Calculate its frequency.

E

i Infrared radiation travels at the same speed as light, but it has a lower frequency than red light. Is its wavelength greater than or less than that of red light?

....

15 Spectra

Definitions to learn

spectrum	waves, or colours of light, separated out in order according to their wavelengths
electromagnetic spectrum	the family of radiations similar to light
infrared radiation	electromagnetic radiation whose wavelength is greater than that of visible light; sometimes known as heat radiation
ultraviolet radiation	electromagnetic radiation whose frequency is higher than that of visible light

Exercise 15.1 Electromagnetic waves

Light waves are just one member of the family of electromagnetic waves.

The visible spectrum is the spectrum of all the colours of light that we can see.

a Which colour in the visible spectrum has the shortest wavelength? ....

b Which colour in the visible spectrum has the highest frequency? ....

c Which colour comes between green and indigo? ....

d Which colour has a wavelength longer than orange light? ....

The diagram represents two waves of visible light, observed for a tiny fraction of a second.

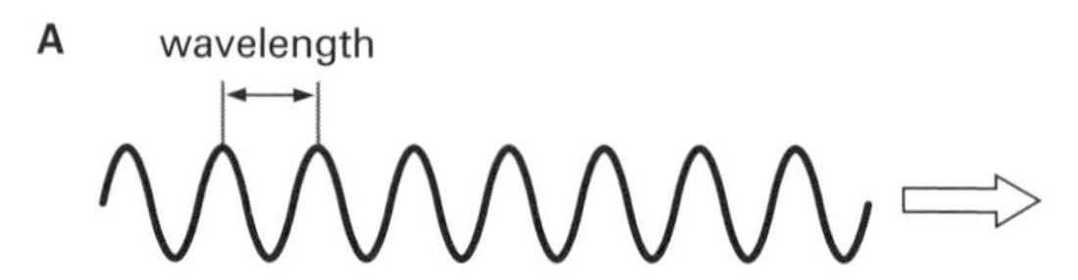

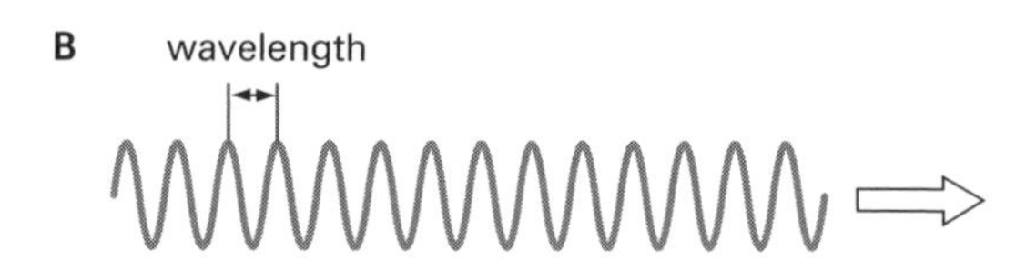

e Which wave (A or B) has the greater wavelength? ....

f How many complete waves are there in A? ....

g How many complete waves are there in B? ....

h How can you tell that the waves are travelling at the same speed?

....

....

i Which wave represents light of a higher frequency? ....

j If the waves represent red and violet light, which one represents red light?

....

The electromagnetic spectrum is the spectrum of all types of electromagnetic radiation, arranged according to their frequencies.

k Which type of electromagnetic radiation has the highest frequency? .

....

l Which type of electromagnetic radiation has the longest wavelength?

....

m Which type of electromagnetic radiation has a frequency just greater than that of

visible light? ....

n Which type of electromagnetic radiation has the most damaging effects on the human

body? ....

Below is a list of types of electromagnetic radiation (on the left) and a list of their uses (on the right). But the lists are not in order!

gamma rays	eyesight
X-rays	transmitting TV programmes
ultraviolet	airport baggage scanners
visible light	cooking food
infrared	sterilising medical equipment
microwaves	communicating with spacecraft
radio waves	tanning skin

o Draw lines to link each type of electromagnetic radiation with its correct use. (There is just one use for each type of radiation.)

Block 4
Electricity and magnetism

16 Magnetism

Definitions to learn

electromagnet	a coil of wire that, when a current flows in it, becomes a magnet
solenoid	a coil of wire that becomes magnetised when a current flows through it
magnetic field	the region of space around a magnet or electric current in which a magnet will feel a force
magnetisation	causing a piece of material to be magnetised; a material is magnetised when it produces a magnetic field around itself
demagnetisation	destroying the magnetisation of a piece of material
hard	describes a material that, once magnetised, is difficult to demagnetise
soft	describes a material that, once magnetised, can easily be demagnetised

Exercise 16.1 Attraction and repulsion

These questions will test your understanding of the attractive and repulsive forces between magnets.

a The diagram below shows two bar magnets. One pole has been labelled. They are repelling each other.

N

- Label the other poles in such a way that the magnets will repel each other.
- Add force arrows to show the magnetic force on each magnet.

b In the next diagram, the two bar magnets are attracting each other.

- Label their poles and add force arrows appropriately.

The diagram on the right shows a horseshoe-shaped permanent magnet attracting a steel rod. The attraction shows that magnetic poles are 'induced' in the rod.

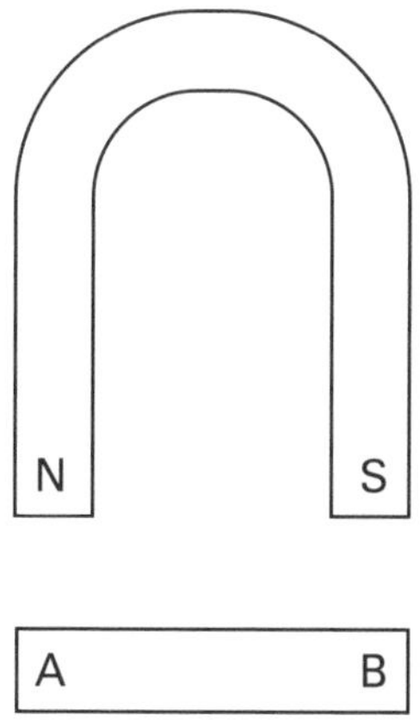

c What type of pole (N or S) must be induced in

end A of the rod? ....

N
S
A
B

d What type of pole (N or S) must be induced in end **B** of the rod?..

The diagram shows a bar magnet that is suspended so that it is free to rotate. It will turn so that its N pole points towards the Earth's geographical North Pole.

e What type of magnetic pole (N or S) must there be close to the Earth's geographical North Pole?

....

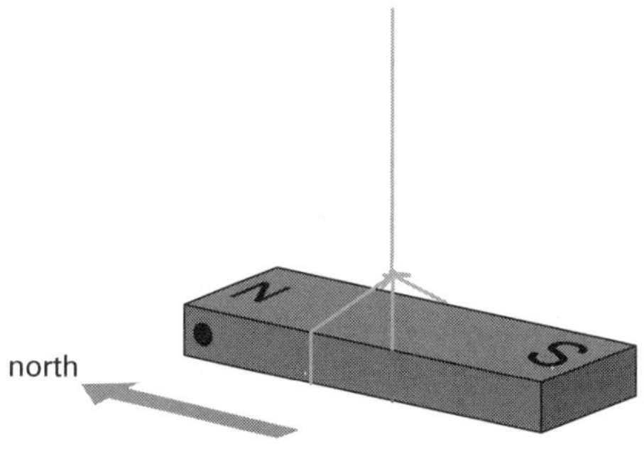

f What type of magnetic pole (N or S) must there be close to the Earth's geographical South Pole?

north
N
S

....

Exercise 16.2 Make a magnet

The process of turning a piece of iron or steel into a magnet is called magnetisation. You could try this at home.

For this exercise, you will need a permanent magnet. A fridge magnet is fine. You will also need a piece of iron or steel to magnetise. A knitting needle or a kitchen skewer may be suitable. You can check by testing that the item is attracted by the magnet.

SAFETY! Take care when working with sharp or pointed objects.

Try to magnetise your steel item by stroking it with the permanent magnet. (This method is described at the bottom of page **171** and in Figure **16.4** on page **172** of the Coursebook.)
After a few strokes, test whether it will attract a paper clip or a staple.
Repeat with a few more strokes (in the same direction). Test again.
Describe a method you can use to tell if the rod is becoming increasingly magnetised.

....

....

Exercise 16.3 Magnetic fields

We use magnetic field lines to represent the shape of a magnetic field. From the pattern, we can also tell if two magnets are attracting each other, or repelling.

Complete the four diagrams below to show the magnetic field around the magnet, around each pair of magnets, and around the electromagnet.

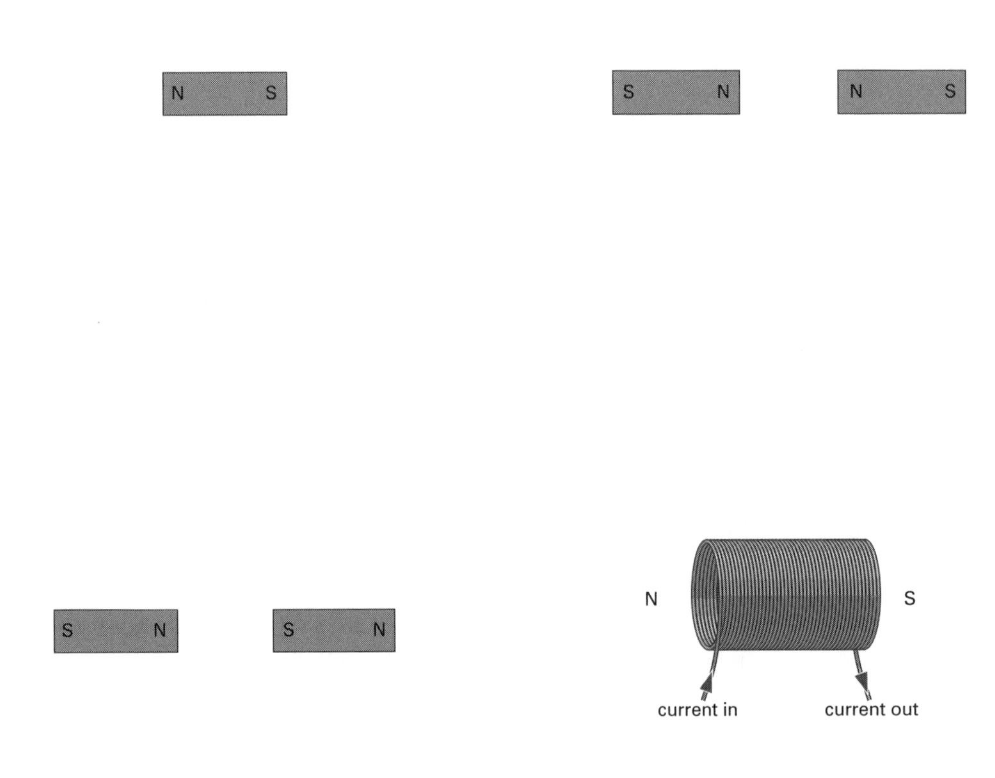

Static electricity

Definitions to learn

electrostatic charge	a property of an object that causes it to attract or repel other objects with charge
positive and negative charge	the two types of electric charge
neutral	having no overall positive or negative electric charge
electric field	a region of space in which an electric charge will feel a force
electron	a negatively charged particle, smaller than an atom
electron charge	the electric charge of a single electron; -1.6×10^{-19} C
proton	a positively charged particle found in the atomic nucleus
proton charge	the electric charge of a single proton; $+1.6 \times 10^{-19}$ C
coulomb (C)	the SI unit of electric charge
induction	a method of giving an object an electric charge without making contact with another charged object

Exercise 17.1 Attraction and repulsion

These questions will test your understanding of the attractive and repulsive forces between electric charges.

a A student rubs a plastic rod with a woollen cloth. The rod and cloth both become electrically charged. What force causes the two materials to become charged?

....

b If the cloth has a positive charge, what type of charge does the rod have?

....

c If the cloth and rod are brought close to one another, will they attract or repel each other? ....

d Explain why this happens.

....

....

e The picture shows one way in which the student could observe the forces exerted by the charged cloth and rod on each other.

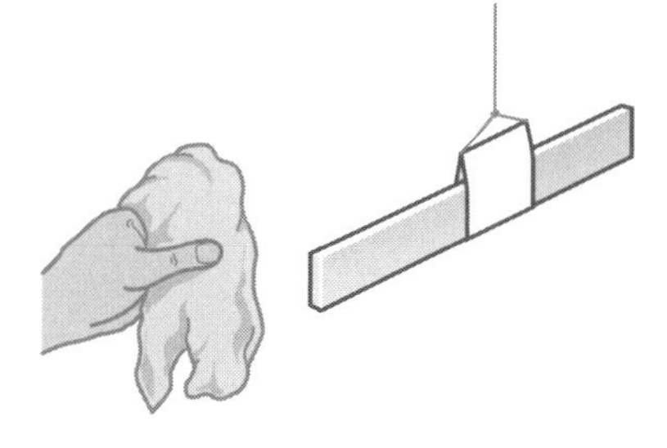

Write a brief description of this experiment – how it is done and what you would expect to observe.

....

....

....

....

....

....

....

....

E

Exercise 17.2 Moving charges

We can understand how an object gains an electric charge by thinking about electrons and protons.

a A student rubs a plastic rod with a woollen cloth. The rod gains a negative electric charge. Before the experiment, the rod had no electric charge. What one word means 'having no electrical charge'? ....

b What type of particles have been transferred to the rod? Explain how you know.

....

....

c The cloth is left with a positive charge. Which type of particle does it have more of, protons or electrons? ....

d A charged object can attract an uncharged object. For example, a charged plastic ruler can attract a small piece of paper. This involves the process of induction. The three diagrams below show how this happens. Under the diagrams on the next page, write a brief explanation of the process.

E

charged ruler

paper

Electrical quantities

Definitions to learn

amp, ampere (A)	the SI unit of electric current
cell	a device that provides a voltage in a circuit by means of a chemical reaction
battery	two or more electrical cells connected together in series; the word may also be used to mean a single cell
conductor	a substance that allows an electric current to pass through it
insulator	a substance that does not conduct electricity
current	the rate at which electric charge flows in a circuit
direct current (d.c.)	electric current that flows in the same direction all the time
alternating current (a.c.)	electric current that flows first one way, then the other, in a circuit

E

coulomb (C)	the SI unit of electric charge; 1 C = 1 A s
ohm (Ω)	the SI unit of electrical resistance; 1 Ω = 1 V/A
p.d. (potential difference)	another name for the voltage between two points
resistance	a measure of the difficulty of making an electric current flow through a device or a component in a circuit
volt (V)	the SI unit of voltage (p.d. or e.m.f.); 1 V = 1 J/C
voltage	the 'push' of a battery or power supply in a circuit

Equations to learn and use

resistance = $\frac{\text{p.d}}{\text{current}}$ $R = \frac{V}{I}$

E

current = $\frac{\text{charge}}{\text{time}}$ $I = \frac{Q}{t}$

power = current × p.d. $P = I \times V$

energy transformed = current × p.d. × time $\Delta E = IVt$

Exercise 18.1 Current and charge

An electric current is a flow of electric charge – the same charge that helps us to explain static electricity.

When there is a current in a circuit, electrons move through the metal wires. The diagram on the right shows a simple circuit in which a cell makes a current flow around the circuit. The arrow shows the direction in which the **electrons** move in the circuit.

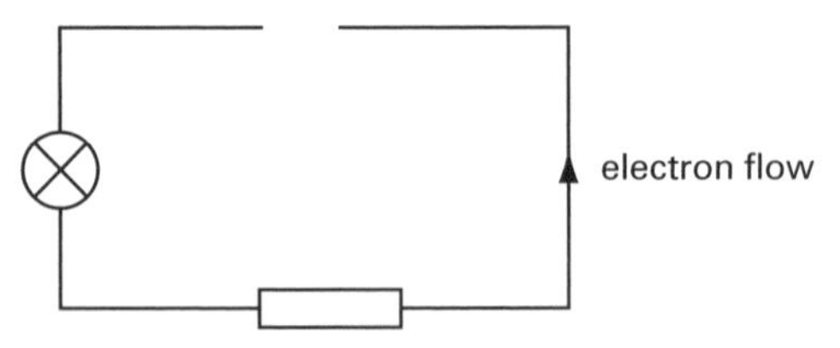

a There is a gap in the circuit where the cell should be. Draw in the symbol for the cell, making sure that it is the correct way round.

b The equation $Q = It$ relates current, charge and time. Complete the table to show the meaning of the symbols in this equation and give their units (name and symbol).

Symbol	Quantity	Unit
Q		
I		
t		

c Write an equation linking the units coulomb, amp and second.

d If a current of 2.4 A flows in a circuit, how much charge flows past a point in 1 s?

....

e In the circuit in part **d**, calculate the charge that flows in 30 s.

f An electric motor is supplied with current by a power supply. If 720 C of charge passes through the motor each minute, what current is flowing?

g A battery supplies a current of 1.25 A to a circuit. How long will it take for 75 C of charge to flow round the circuit?

Exercise 18.2 Electrical resistance

The resistance of a component tells us how easy (or difficult) it is to make current flow through it.

We say that an ohm (Ω) is a volt per amp. So, if a resistor has a resistance of 10 Ω, it takes 10 V to make a current of 1 A flow through it.

a What voltage is needed to make a current of 2 A flow through the same 10 Ω resistor?

....

b What voltage is needed to make a current of 1 A flow through a 20 Ω resistor?

....

c The current in a circuit changes as the resistance in the circuit changes. Complete the table below by indicating whether the change indicated will cause the current to increase or decrease.

Change	Current – increase or decrease?
more resistance in the circuit	
less resistance in the circuit	
increase the voltage	
use thinner wires	
use longer wires	

d Use the equation $R = \frac{V}{I}$ to calculate the resistance of a lamp if a p.d. of 36 V makes a current of 4.5 A flow through it.

e A student measured the resistance of a resistor. To do this, she set up a circuit in which the resistor was connected to a variable power supply, a voltmeter and an ammeter. In the space below, draw a circuit diagram to represent these components connected together correctly so that the student could measure the current in the resistor and the p.d. across it.

f The table below shows the student's results. Complete the third column.

P.d. V / V	Current I / A	Resistance R / Ω
2.0	0.37	5.4
4.1	0.75	
5.9	1.20	
7.9	1.60	

g Calculate an average value for the resistance R of the resistor.

Average value for the resistance R = ...

E

Exercise 18.3 Electrical energy and power

Power has the same meaning in electricity as it had when we were considering forces. It is the rate at which energy is transferred – in this case, by an electric current.

a Write down an equation linking power, energy transformed and time.

b Write down an equation linking power, current and p.d.

c An electric motor is connected to a 12 V supply. A current of 0.25 A flows through the motor. Calculate the power of the motor.

An electrical appliance has a label that indicates its power. The label includes the following data:

110 V   500 W   50 Hz

d What is the power rating of the appliance? ....

e How much energy does it transform each second? ....

f How can you tell that the appliance works with alternating current?

....

g Calculate the current that will flow when the appliance is in normal use.

E

h A lamp has a resistance of 600 Ω. Calculate the current that flows through the lamp when it is connected to a 240 V mains supply.

i Calculate the power of the lamp in part **h**.

19 Electric circuits

Definitions to learn

variable resistor	a resistor whose resistance can be changed, for example by turning a knob
thermistor	a resistor whose resistance changes a lot over a small temperature range
potential divider	a part of a circuit consisting of two resistors connected in series
light-dependent resistor (LDR)	a device whose resistance decreases when light shines on it
diode	an electrical component that allows electric current to flow in one direction only
light-emitting diode (LED)	a type of diode that emits light when a current flows through it
capacitor	a device used for storing energy in an electric circuit
relay	an electromagnetically operated switch
fuse	a device used to prevent excessive currents flowing in a circuit
circuit breaker	a safety device that automatically switches off a circuit when the current becomes too high
transistor	an electrical component that can act as an electronic switch
logic gate	an electronic component whose output voltage depends on the input voltage(s)
truth table	a way of summarising the operation of a combination of logic gates

Equations to learn and use

Resistors in series: $R = R_1 + R_2 + R_3$

Resistors in parallel: $\frac{1}{R} = \frac{1}{R_1} + \frac{1}{R_2} + \frac{1}{R_3}$

Exercise 19.1 Circuit components and their symbols

In the previous chapter, you made use of a few electrical circuit symbols. How many others do you know? What is the function of each component?

a Complete the table below by drawing in the symbol for each component.

lamp	resistor	switch
LDR	thermistor	fuse
diode	capacitor	transformer

b Complete the table below by identifying each component described in the first column. (The component names are all in part **a** above.)

Description	Component
gives out heat and light	
resistance changes as the temperature changes	
stores energy in a circuit	
'blows' when the current is too high	
allows current to flow one way only	
makes and breaks a circuit	
has less resistance on a sunny day	
used to change the voltage of alternating current	

E

Exercise 19.2 Diodes

A diode will only let current flow in one direction – in the direction of the arrow.

In the circuits below, there are several lamps and light-emitting diodes (LEDs).
For each LED, decide whether it will light up. Mark each LED that lights up with a tick.
For each lamp, decide whether or not it will light up. Mark each lamp that lights up with a tick.

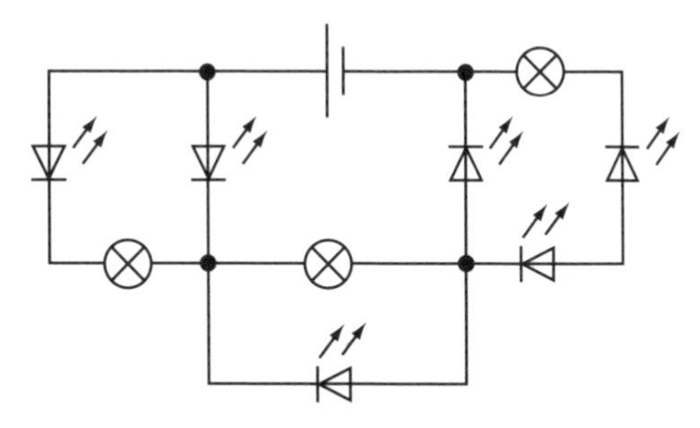

Exercise 19.3 Resistor combinations

Test your understanding of how current flows in a circuit with more than one resistor.

a Calculate the combined resistance of four 120 Ω resistors connected in series.

b In the circuit shown below, are the three resistors connected in series or in parallel?

....

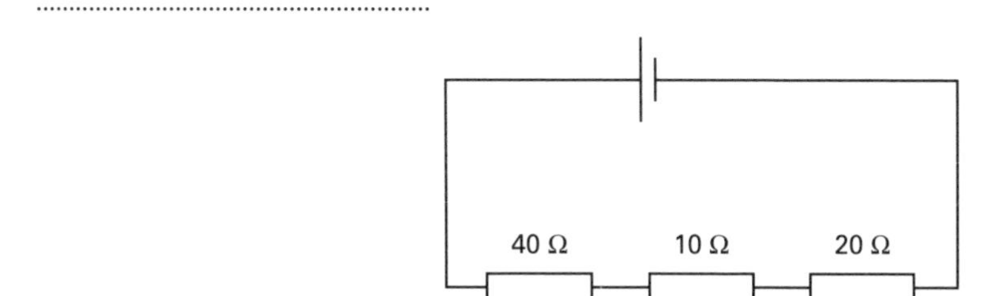

c Calculate the combined resistance of the three resistors in the circuit in part **b** above.

d What can you say about the current in the circuit in part **b** above?

....

e In the circuit shown below, are the two resistors connected in series or in parallel?

....

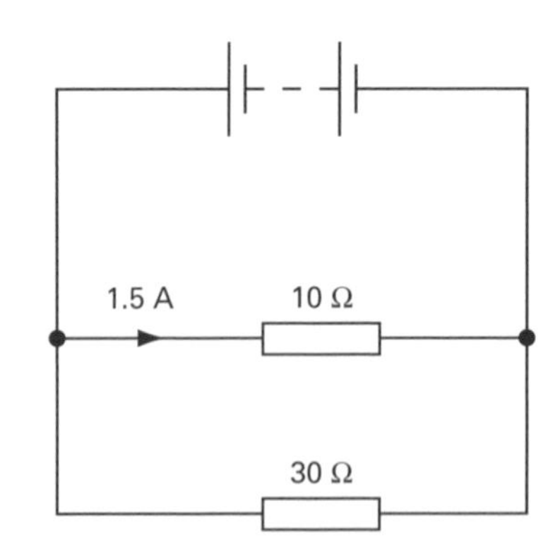

f Which of the following statements is true? (Tick the correct statement.)

- The combined resistance of the two resistors must be less than 10 Ω.
- The combined resistance of the two resistors must be more than 40 Ω.

g The diagram in part **e** above shows the current in the 10 Ω resistor. Calculate the current in the 30 Ω resistor.

Electromagnetic forces

Definitions to learn

electromagnet	a coil of wire that, when a current flows in it, becomes a magnet
corkscrew rule	the rule used to determine the direction of the magnetic field around an electric current
relay	an electromagnetically operated switch
commutator	a device used to allow current to flow to and from the coil of a d.c. motor or generator
Fleming's left-hand rule	a rule that gives the relationship between the directions of force, field and current when a current flows across a magnetic field
cathode ray	a ray of electrons travelling from cathode to anode in a vacuum tube
thermionic emission	the process by which cathode rays (electrons) are released from the heated cathode of a cathode-ray tube

E

Exercise 20.1 Using electromagnetism

Every time an electric current flows, a magnetic field appears around it. We make use of this effect in a number of devices.

The direction of the magnetic field lines around a current is given by the 'pencil-sharpener rule' (or 'corkscrew rule'). Complete the following sentences:

a The direction in which you push the pencil into the sharpener tells you the direction of

....

b The direction in which you turn the pencil tells you the direction of

....

A relay is an electromagnetically operated switch. The diagram below shows a circuit that makes use of a relay.

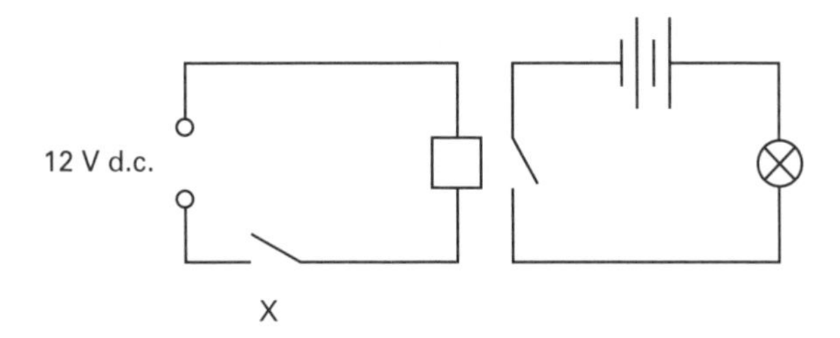

c Copy the circuit symbol for the relay into the space below. Label the parts of it that represent the coil and the switch.

d State and explain what happens when switch **X** in the circuit is closed.

....

....

....

....

....

....

e An electric motor can be made using a coil of wire that rotates in a magnetic field.

Which part of the motor acts as an electromagnet? ....

f Electric current enters and leaves the coil via two brushes. Name the part of the motor against which the brushes press, to transfer the current to the coil.

....

The apparatus shown in the diagram below is used to demonstrate the force on a current-carrying conductor in a magnetic field.

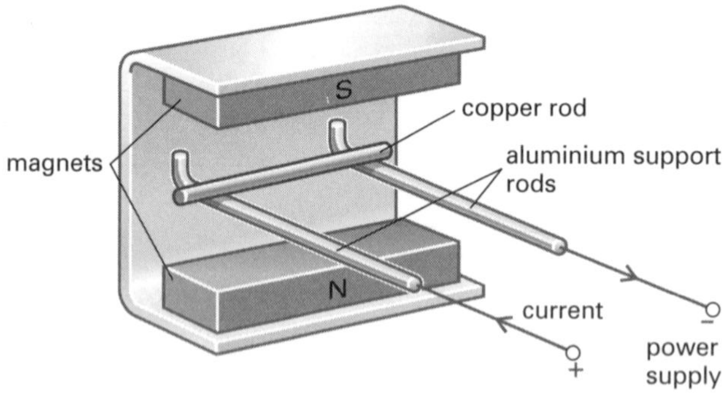

S
copper rod
magnets
aluminium support rods
N
current
+
−
power supply

g On the diagram, draw an arrow to show the direction of the magnetic field.

h In this arrangement, the force on the copper rod will make it roll towards the power supply. What effect would reversing the direction of the current have?

....

....

i State **two** ways in which the force on the copper rod could be increased.

....

....

Exercise 20.2 Cathode rays

Cathode rays are used in some (now mainly older) television tubes. When physicists first discovered them, cathode rays revealed a lot about some of the fundamental particles of nature.

The diagram shows a cathode-ray tube, which is fitted with two pairs of plates that can be used to control the direction of the beam.

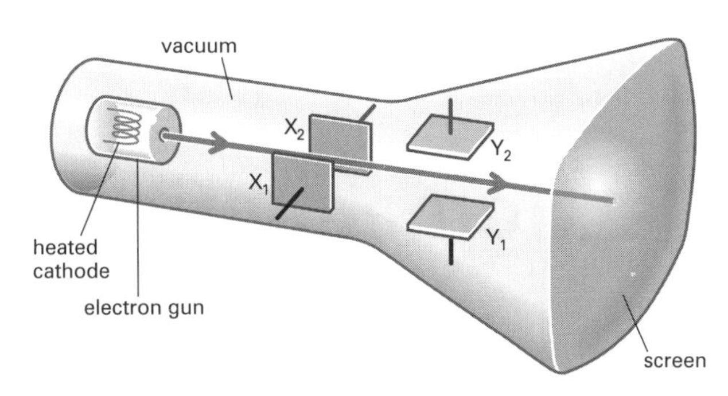

a What particles are emitted from the cathode to create the beam of cathode rays?

....

b What electric charge do these have, positive or negative? ....

c Name the process by which they are emitted. ....

d To deflect the beam upwards, plate Y_1 must be given an electric charge. Should this be positive or negative? ....

e Plate Y_2 must also be given an electric charge. Should this be positive or negative?

....

f What is the purpose of plates X_1 and X_2?

....

....

g If a varying voltage is connected across plates Y_1 and Y_2, what will be observed on the screen?

....

....

h Why must there be a vacuum in the tube?

....

....

....

21 Electromagnetic induction

Definitions to learn

dynamo effect electricity is generated when a coil moves near a magnet
a.c. generator a device, such as a dynamo, used to generate alternating current (a.c.)
national grid the system of power lines, pylons and transformers used to carry electricity around a country
power lines cables used to carry electricity from power stations to consumers
slip rings a device used to allow current to flow to and from the coil of an a.c. motor or generator
transformer a device used to change the voltage of an a.c. electricity supply

Equations to learn and use

$$\frac{\text{voltage across primary coil}}{\text{voltage across secondary coil}} = \frac{\text{number of turns on primary}}{\text{number of turns on secondary}}$$

$$\frac{V_P}{V_S} = \frac{N_P}{N_S}$$

E

power into primary coil = power out of secondary coil

$$I_p \times V_p = I_s \times V_s$$

Exercise 21.1 Electricity generation

Electromagnetic induction is the process in which a current is made to flow when a conductor moves in a magnetic field.

a Complete the table by indicating whether a current will be induced (made to flow). In each case, put 'Yes' or 'No' in the second column. (You can assume that the wire is part of a complete circuit.)

Case	Current induced?
a wire is moved through the field of a magnet	
a magnet is held close to a wire	
a magnet is moved into a coil of wire	
a magnet is moved out of a coil of wire	
a magnet rests in a coil of wire	

b Alternating current is generated using an a.c. generator. This is similar to an electric motor, working in reverse. An a.c. generator does not have a commutator. Instead, current enters and leaves the spinning coil through brushes. which press on the

....

c The diagram shows how an alternating current varies with time. On the diagram, mark one cycle of the alternating current.

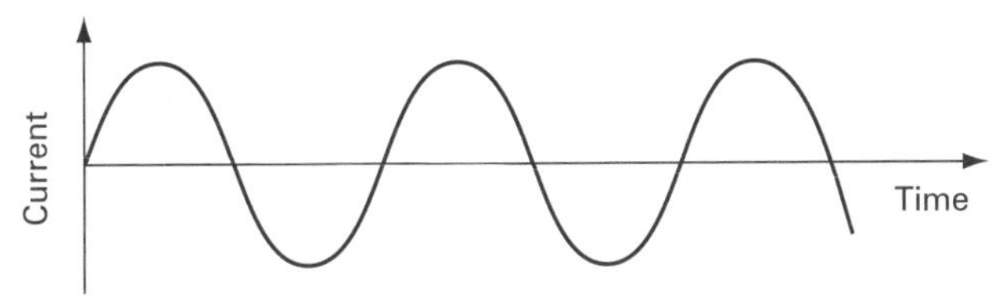

d In a power station, high-pressure steam from the boilers causes a ....

to spin, and this makes the generators turn. The voltage of the electricity coming from

the generators is increased using .... The electricity is

then distributed to consumers via the power lines of the ....

Exercise 21.2 Transformers

Transformers make use of electromagnetic induction to change the voltage and current of an alternating supply.

a The diagram shows a step-up transformer. In such a transformer, which coil has more turns, the primary or the secondary? ....

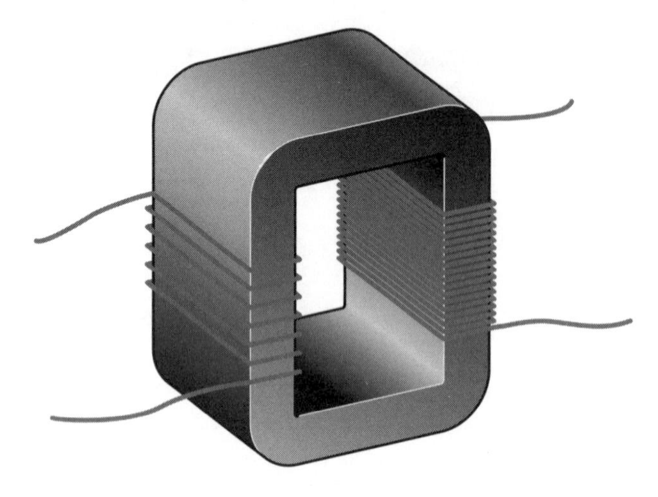

b On the diagram above, label the primary coil, the secondary coil and the iron core.

In the diagram below, a transformer is being used to change the mains voltage to a lower value so that it will light a 12 V lamp.

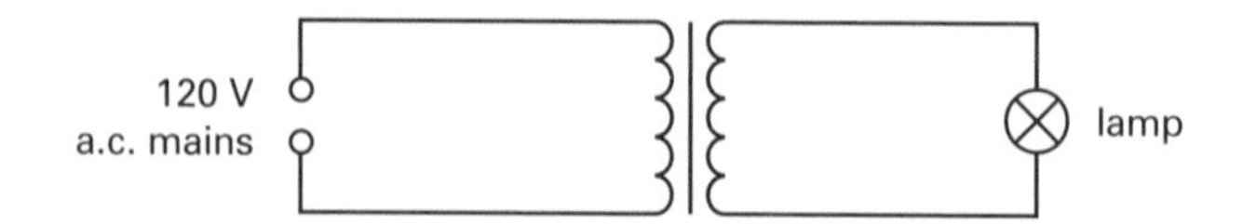

c Is this a step-up or step-down transformer? ....

d If the primary coil has 1000 turns, how many turns must the secondary have?

E

e At a small power station, the generator produces alternating current at a voltage of 10 kV. This must be reduced to 415 V for use in a factory. The transformer used for this purpose has a primary coil of 2000 turns. How many turns must its secondary coil have?

f In normal operation, the current flowing from the generator in part **e** is 4.5 A. What power is being generated?

g Calculate the current flowing in the cables in the factory. (Assume that all of the electrical power generated is transmitted to the factory.)

Block 5

Atomic physics

22 The nuclear atom

Definitions to learn

electron	a negatively charged particle, smaller than an atom
proton	a positively charged particle found in the atomic nucleus
proton number (*Z*)	the number of protons in an atomic nucleus
neutron	an electrically neutral particle found in the atomic nucleus
neutron number (*N*)	the number of neutrons in the nucleus of an atom
nucleon	a particle found in the atomic nucleus: a proton or a neutron
nucleon number (*A*)	the number of protons and neutrons in an atomic nucleus
nuclide	a 'species' of nucleus having particular values of proton number and nucleon number
isotope	isotopes of an element have the same proton number but different nucleon numbers

E

An equation to learn and use

E

proton number + neutron number = nucleon number $\quad Z + N = A$

Exercise 22.1 The structure of the atom

Everything is made of atoms, but what are atoms made of?

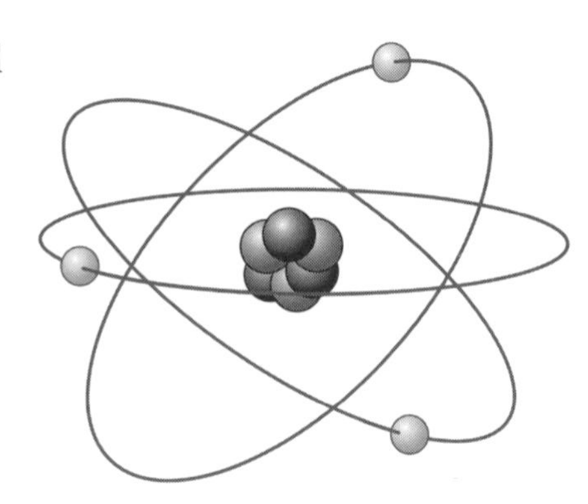

a The diagram shows a simple model of an atom. Label the nucleus and an electron.

b Which part of the atom contains most of its mass?

....

c Which part of the atom contains all of its positive charge? ....

d Complete the table by identifying the particles described.

Description	Protons / neutrons / electrons?
These particles make up the nucleus	
These particles orbit the nucleus	
These particles have very little mass	
These particles have no electric charge	
These particles have an exactly opposite charge to an electron	

e The nucleus of a particular atom of carbon (C) is represented like this: ${}^{13}_{6}\text{C}$

State the value of its proton number *Z*. ....

State the value of its nucleon number *A*. ....

Calculate the value of its neutron number *N*.

f The nucleus of a particular atom of oxygen (O) is made up of 8 protons and 8 neutrons.

Write the symbol for this nucleus in the form ${}^{A}_{Z}\text{X}$. ....

E

Exercise 22.2 Discovering the structure of the atom

How do we know that an atom has a nucleus with electrons in orbit around it?

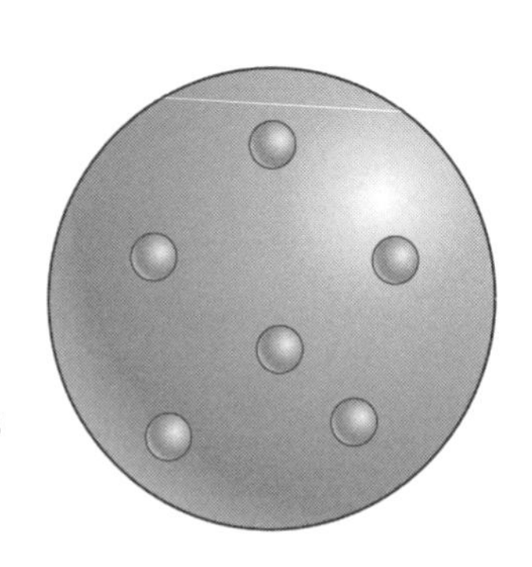

a The diagram on the right shows J. J. Thomson's model of the atom. Write a brief description of his model, and label its parts in the diagram.

....

....

E

b Rutherford used alpha radiation to investigate atoms in a thin sheet of gold foil. Cross out the incorrect words in the following sentences:

- Alpha particles have a **positive / negative** charge.
- An alpha particle is **bigger / smaller** than a gold atom.
- The nucleus of a gold atom has a **positive / negative** charge.

The diagram shows the paths of some alpha particles as they approached the nucleus of a gold atom.

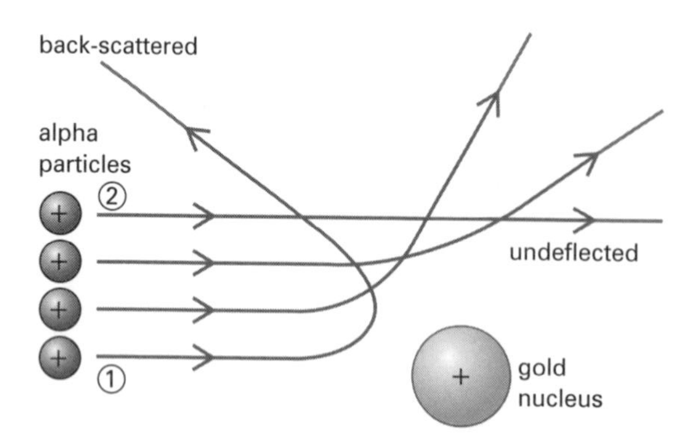

c Explain what it means to say that an alpha particle was 'back-scattered'.

....

....

....

....

d Explain why the alpha particle (1) shown was back-scattered.

....

....

....

e Explain why the alpha particle (2) shown was undeflected as it passed the gold nucleus.

....

....

....

E

Exercise 22.3 Isotopes

Atoms of an element come in more than one form. These different forms are called isotopes.

a What is the same for two isotopes of an element?

....

....

b What is different for two isotopes of an element?

....

....

c The diagram below represents an atom of an isotope of boron (B). Write the symbol for this nuclide in the form ${}^{A}_{Z}\text{X}$. ....

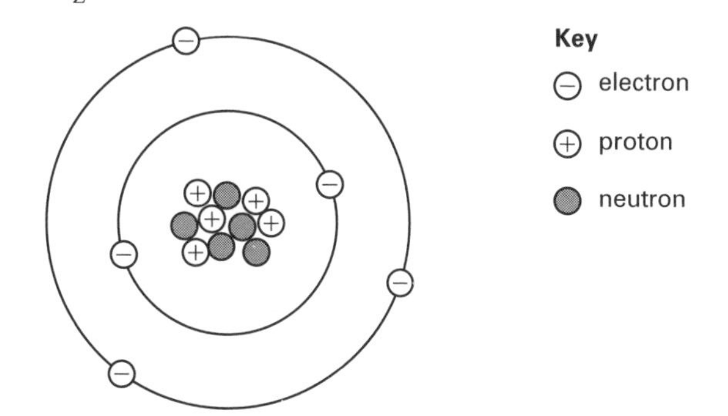

d The table below shows some values of *Z*, *N* and *A* for six different isotopes. Complete the table as follows. Fill in the missing values of *Z*, *N* and *A*. Use a Periodic Table to identify the elements. Then, in the last column, write the symbol for each nuclide in the form ${}^{A}_{Z}\text{X}$.

Isotope	Proton number *Z*	Neutron number *N*	Nucleon number *A*	Name of element	Nuclide symbol ${}^{A}_{Z}\text{X}$
I-1	4	5			
I-2	5	7			
I-3		4	8		
I-4	6		11		
I-5		6	11		

Radioactivity

Definitions to learn

radioactive substance a substance that decays by emitting radiation from its atomic nuclei

radiation energy spreading out from a source carried by particles or waves

background radiation the radiation from the environment to which we are exposed all the time

contaminated when an object has acquired some unwanted radioactive substance

irradiated when an object has been exposed to radiation

penetrating power how far radiation can penetrate into different materials

radioactive decay the decay of a radioactive substance when its atomic nuclei emit radiation

alpha decay the decay of a radioactive nucleus by emission of an alpha particle

alpha particle (α-particle) a particle of two protons and two neutrons (a helium nucleus) emitted by an atomic nucleus during radioactive decay

beta decay the decay of a radioactive nucleus by emission of a beta particle

beta particle (β-particle) a particle (an electron) emitted by an atomic nucleus during radioactive decay

gamma ray (γ-ray) electromagnetic radiation emitted by an atomic nucleus during radioactive decay

radioisotope a radioactive isotope of an element

ionisation when a particle (atom or molecule) becomes electrically charged by losing or gaining electrons

ionising radiation radiation, for example from radioactive substances, that causes ionisation

random process a process that happens at random rather than at a steady rate; in radioactive decay, it is impossible to predict which atom will be the next to decay, or when a given atom will decay

half-life the average time taken for half the atoms in a sample of a radioactive material to decay

radioactive tracing a technique that uses a radioactive substance to trace the flow of liquid or gas, or to find the position of cancerous tissue in the body

radiocarbon dating a technique that uses the known rate of decay of radioactive carbon-14 to find the approximate age of an object made from dead organic material

Exercise 23.1 The nature of radiation

Radioactive substances emit radiation. As it passes through a material, it may be absorbed. This helps us to distinguish the three types of radiation.

a The diagram shows how the three types of radiation from radioactive substances are absorbed by different materials.

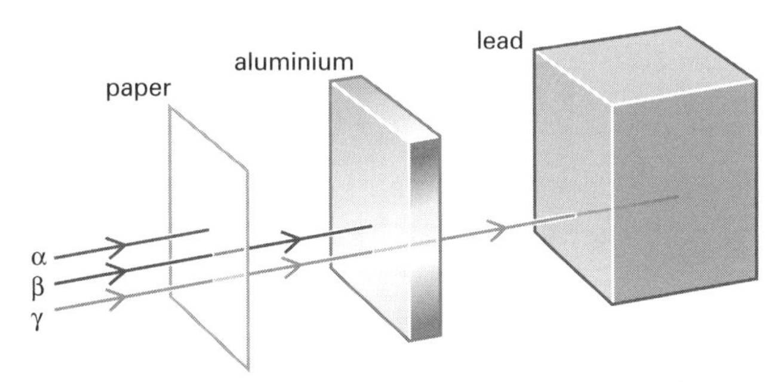

The diagram uses symbols – write the full names here:

α ....

β ....

γ ....

b Which type of radiation is the most penetrating? ....

c Which type of radiation can be absorbed by a few centimetres of air or by a thin sheet of paper? ....

d Which types of radiation are absorbed by a thick sheet of lead? ....

e The radiation from radioactive substances is called 'ionising radiation'. This is because it can change particles, causing them to become ions. What is an ion?

....

....

f Which type of ionising radiation has no mass? ....

g Which type of ionising radiation has a positive charge? ....

h Which type of ionising radiation is an electron? ....

i Which type of ionising radiation is the same as a helium nucleus? ....

j Which type of ionising radiation travels at the speed of light? ....

k Which type of ionising radiation has a negative charge? ....

l Which type of ionising radiation is a form of electromagnetic radiation?

....

Exercise 23.2 Radioactive decay

The decay of radioactive substances follows a particular pattern, which arises from the random nature of decay.

a A sample of a radioactive substance contains 2400 undecayed atoms. Calculate the number that will remain after three half-lives.

b For the sample in part **a**, calculate the number that will decay during three half-lives.

c A sample of a radioactive substance contains 1000 undecayed atoms. Its half-life is 4.5 years. Calculate the number that will remain undecayed after 9.0 years.

d A radioactive substance has a half-life of 13 years. Calculate the time it will take for the number of undecayed atoms in a sample to fall to one-eighth of their original number.

e The table shows how the activity of a radioactive sample changed as it decayed.

Time / h	0	2	4	6	8
Activity / counts per second	500	280	160	95	55

On the grid below, draw a graph of activity against time and use it to deduce the half-life of the substance. Show your method on the graph.

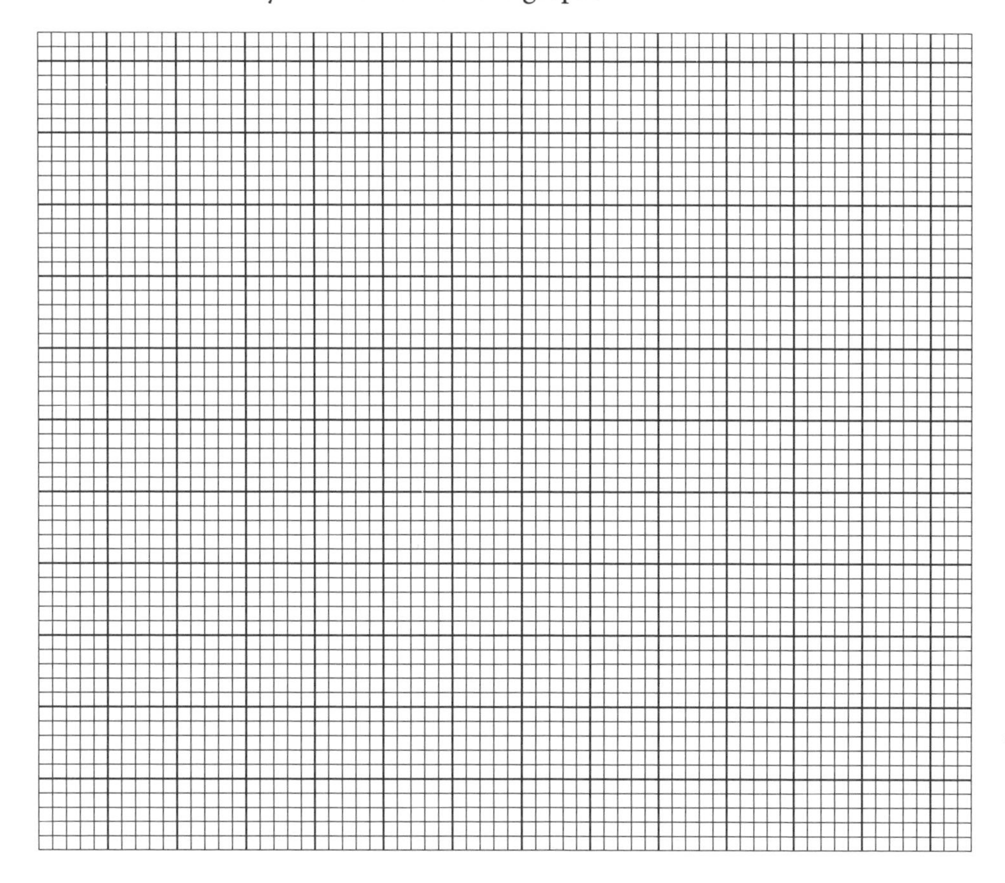

Half-life is approximately ....

The graph below shows the amount of undecayed material in a sample of a radioactive substance as it decayed. When the material had decayed to a very low level, the detector still recorded background radiation.

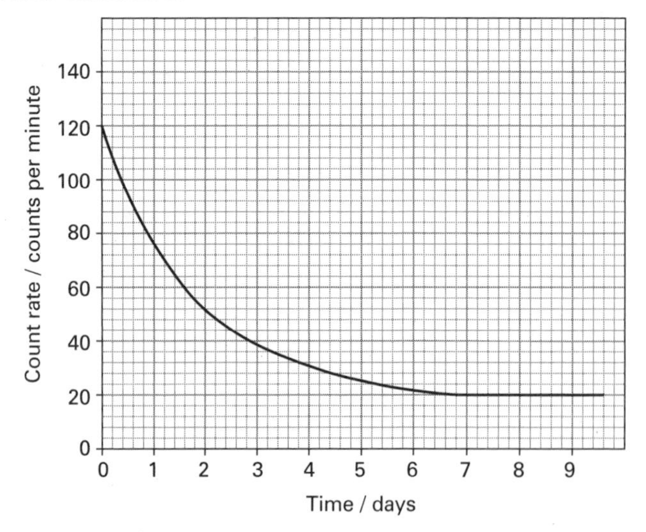

f From the graph, determine the background count rate. ....

g Determine the initial count rate due to the radioactive substance by calculating:

initial count rate – background count rate

h Determine the approximate half-life of the substance. There are **two** different ways to do this. Can you find both? Show your working below, and on the graph.

Approximate half-life is ....

E

Exercise 23.3 Using radioactive substances

Radioactive substances are useful. In particular, we can use the radiation they produce. For this, it is important to understand the properties of ionising radiation.

Complete the table below as follows.

- The first column shows some uses of radioactive substances.
- In the second column, write the appropriate code number or numbers from the list below:

  1 Some radiation is very penetrating

  2 Some radiation is readily absorbed

  3 Ionising radiation damages cells

  4 Radiation is easily detected

  5 Radioactive substances decay at a known rate

Use of radioactivity	Code number(s)
Finding the age of an ancient object	
Destroying cancerous tissue	
Imaging a tumour in the body	
Sterilising medical equipment	
Controlling the thickness of paper in a paper mill	
Detecting smoke in the air	
Tracing leaks from underground pipes	